# HAZOP: Guide to Best Practice

Guidelines to Best Practice for the Process and Chemical Industries

# HAZOP: Guide to Best Practice
## Guidelines to Best Practice for the Process and Chemical Industries

### Third Edition

**Frank Crawley**
Atkins, University of Strathclyde

**Brian Tyler**
S&T Consultants

*(based upon the earlier editions by
Frank Crawley, Malcolm Preston, and Brian Tyler)*

**ELSEVIER**

AMSTERDAM • BOSTON • HEIDELBERG • LONDON • NEW YORK • OXFORD
PARIS • SAN DIEGO • SAN FRANCISCO • SINGAPORE • SYDNEY • TOKYO

Elsevier
Radarweg 29, PO Box 211, 1000 AE Amsterdam, Netherlands
The Boulevard, Langford Lane, Kidlington, Oxford OX5 1GB, UK
225 Wyman Street, Waltham, MA 02451, USA

Third Edition 2015

ISBN: 978-0-323-39460-4

**Library of Congress Cataloging-in-Publication Data**
A catalog record for this book is available from the Library of Congress

**British Library Cataloguing-in-Publication Data**
A catalogue record for this book is available from the British Library

For Information on all Elsevier publications
visit our website at http://store.elsevier.com/

This book has been manufactured using Print On Demand technology.

Working together
to grow libraries in
developing countries

www.elsevier.com • www.bookaid.org

# CONTENTS

# FOREWORD

## FOREWORD TO THIRD EDITION

It is with great pleasure that I have been invited to offer a preface to this, the third edition of *HAZOP Guide to Best Practice* which is certainly one of the most popular IChemE texts that has been developed by EPSC members since the inception of the Centre in 1992.

This particular book has fond personal memories because several years ago when working in industry I attended an IChemE HAZOP for Team Leaders course with one of its authors, Brian Tyler, which was held at the former UMIST campus in Manchester. The opportunity presented during that course of managing a HAZOP study team gave me the necessary confidence back at the workplace to train frontline staff in the use of the technique, then lead a team in a study on a fully functioning gin distillery, and finally present the recommendations to the site executive team. I still have the course folder, and I am glad to see that much of that content still forms the core of this current text.

Nevertheless much has happened since the first edition and nothing stands still for long and so for the technique of HAZOP. There is now much greater appreciation in HAZOP studies of human error in accident causation and more broadly human factors and the role of automation. HAZOP studies are now performed routinely on continuous plants at various stages of operation such as start-up and shutdown, batch processing plants, and even packaging plants. The technique of deviation analysis inherent in the method lends itself with care and imagination to many diverse major hazard environments.

As for this third edition, the authors, Brian Tyler and Frank Crawley, are to be praised for their collective efforts in revising this book yet again and keeping the content as fresh and topical as possible. HAZOP provides both a structure for the team identification of hazards, accident scenarios, and operability issues while offering the chance for an element of creative thinking for a team whose time and effort is well managed. If anything the pressure in recent years has grown on the typical HAZOP team to identify and address all conceivable hazards arising from dangerous operations which makes this new edition particularly welcome for students and practitioners alike.

I have no doubt that EPSC members who contributed to the first edition will be immensely proud to see that this book has become a standard reference among the process safety community.

**Lee Allford**
EPSC Operations Manager
IChemE
Davis Building
Rugby CV21 3HQ
December 2014

## FOREWORD TO EARLIER EDITIONS

Hazard and Operability Studies (later shortened to *HAZOP*) were devised by ICI in the late 1960s, following some major problems with new, large process plants. The study was an evolution of method study and was used during the design stage of a project to identify and correct design faults which might lead to Hazard or Operability problems. Over the last few decades, the need for high standards in safety and the environment is fully recognized by the Regulator, the industry, and the public.

HAZOP is now the first choice tool for the identification of weaknesses in the process design and is used worldwide within the process industry. It has been used in a modified form outside the process industry.

The first definitive guide on Hazard and Operability Studies was issued by the Chemical Industries Association in 1974 when the tool had been fully developed. This remained the main guidance for 26 years. However, in that period, new ideas on HAZOPs had been developed and equally some poor practices had been adopted. In 1998, it was decided that a new guide, using best practice, should be written. The first edition of this guide was published in 2000. For the second edition, the authors took the opportunity of reviewing that guide and incorporating better practices and giving more guidance on how these might be applied. In particular the new guide addresses Computer-Controlled Processes.

Although the basic approach of HAZOP is unchanged, there is now considerable experience in how the technique can be used most effectively. This experience has been drawn upon in preparing this guide, with a total of 31 companies contributing to its preparation.

Finally, the guide is important as a joint project which has the support of the Institution of Chemical Engineers (IChemE), the Chemical Industries Association (CIA), and the European Process Safety Centre (EPSC).

Our thanks are due both to the authors of the guide and to the many industrial members who assisted in its development.

**Richard Gowland**
Technical Director
European Process Safety Centre

# ACKNOWLEDGMENTS

This first edition of this guide would not have been possible without the contributions of the following individuals and the support of their companies:

| | |
|---|---|
| M. Dennehy | Air Products plc |
| J. Snadden | Air Products plc |
| A Jacobsson | AJ Risk Engineering AB |
| D. Hackney | Akcros Chemicals |
| K. Brookes | Akcros Chemicals |
| R. Vis van Heemst | Akzo Nobel Engineering bv |
| K. Gieselburg | BASF AG |
| K-O. Falke | Bayer AG |
| V. Pilz | Bayer AG |
| H. Crowther | BP Chemicals |
| H. Jenkins | BP Chemicals |
| J-H. Christiansen | Borealis |
| M. Scanlon | Chemical Industries Association |
| F. Altorfer | Ciba Specialty Chemicals |
| T. Gillard | Consultant |
| S. Duffield | EC JRC |
| J-C. Adrian | Elf Atochem |
| P. Lawrie | EniChem UK Ltd |
| M. Powell-Price | EPSC |
| R. Turney | EPSC |
| R. Carter | Eutech |
| A. Ormond | Eutech |
| C. Swann | Eutech |
| S. Turner | Foster Wheeler |
| H. Wakeling | Foster Wheeler |
| T. Maddison | Health and Safety Executive (UK) |
| A. Rushton | Health and Safety Executive (UK) |
| K. Patterson | Hickson and Welch |
| J. Hopper | Huntsman Polyurethanes |
| M. Wilkinson | Institution of Chemical Engineers |

| J. Geerinck | Janssen Pharmaceutica NV |
| G. Gillett | Kvaerner Water |
| A. Poot | Lyondell Chemie Nederland Ltd |
| E. Dyke | Merck |
| C. Downie | Merck and Co Ltd |
| K. Dekker | Montell Polyolefins bv |
| J. Ytreeide | Norsk Hydro AS |
| D. Moppett | Proctor and Gamble European Technical Centre |
| P. Rouyer | Rho̅ ne-Poulenc |
| C. Caputo | Rohm and Haas Italia |
| G. Dilley | Rohm and Hass (Scotland) Ltd |
| F. de Luca | Snamprogetti S.p.A |
| C. Bartholome̅ | Solvay SA |
| P. Depret | Solvay SA |
| J. Ham | TNO |
| K. Ling | Zurich International |

The initiative of the IChemE Safety and Loss Prevention Subject Group and the support of the IChemE Safety Health and Environmental Policy Committee are recognized.

While all of the above contributed to this guide, the authors would like to give special thanks to the following for their contributions and support: Rob Carter, Howard Crowther, Martin Dennehy, Julian Hooper, Hedley Jenkins, Ken Patterson, Chris Swann, Simon Turner, and Robin Turney.

*The authors would also like to give special thanks to Phil Aspinall of VECTRA Group Limited for his help and contribution in preparing the second edition of this guide, in particular for the development of Section 10.1 on HAZOP Study of Computer-Controlled Processes.*

*The authors of the third edition wish to thank Jerry Lane (Optimus) for his help during the preparation of this revision.*

# CHAPTER 1

# Introduction

## 1.1 AIMS AND OBJECTIVES

This book is intended to provide guidance on a specific technique developed for use in the process and chemical industries. The technique described is HAZOP (hazard and operability) study, a detailed method for systematic examination of a well-defined process or operation, either planned or existing.

The HAZOP study method was developed by ICI in the 1960s and its use and development was encouraged by the Chemical Industries Association (CIA) Guide published in 1977. Since then, it has become the technique of choice for many of those involved in the design of new processes and operations. In addition to its power in identifying safety, health, and environmental (SHE) hazards, a HAZOP study can also be used to search for potential operating problems. Not surprisingly, the method has been applied in many different ways within the process industries.[1]

While it is frequently used on new facilities, it is now often applied to existing facilities and modifications. It has also been successfully applied to process documentation, pilot plant, and hazardous laboratory operations as well as tasks such as commissioning and decommissioning, emergency operations, and incident investigation.

The objective here is to describe and illustrate the HAZOP study method, showing a variety of uses and some of the approaches that have been successful within the process industry. An important input has come from European Process Safety Centre (EPSC) members where 22 member companies responded in a survey carried out prior to the first edition of this Guide (2000). This identified many features generally regarded as essential to a good study. In addition, many common variations were described. These variations are in part due to the range of problems encountered within industry but also reflect individual choices of style. HAZOP study is a versatile technique and

HAZOP: Guide to Best Practice. DOI: http://dx.doi.org/10.1016/B978-0-323-39460-4.00001-3

good results may be achieved by several different approaches provided the basic principles are followed. It is hoped that this Guide will help maintain a high standard for HAZOP study within the industry, both by raising quality and encouraging flexibility without putting any unnecessary constraints upon its use and development.

The HAZOP study method is well developed and is useful in most applications. There are other methods, however, that may have to be considered depending on the complexity and hazards of the installation being constructed and the state of the design. This publication does not address these methods in detail but their importance is discussed in Chapter 2. A fuller account is given in the IChemE Guide, Hazard Identification Methods.[2]

Finally, three illustrations of process industry applications are given in Appendices 3, 4 and 5. These examples cannot fully represent all the possible applications and process industries and readers new to HAZOP study are advised to consult the reference list,[3-6] the International Electrotechnical Commission (IEC) standard,[7] or guidelines written specifically for other industries.

It is hoped that this guide will help people within the process industries, including all those with responsibilities within safety management systems. Although it is primarily written for HAZOP study leaders, scribes, and members, it may also be of use to those involved in training and plant management.

## 1.2 ESSENTIAL FEATURES OF HAZOP STUDY

A HAZOP study is a structured analysis of a system, process, or operation for which detailed design information is available, carried out by a multidisciplinary team. The team proceeds on a line-by-line or stage-by-stage examination of a firm design for the process or operation. While being systematic and rigorous, the analysis also aims to be open and creative. This is done by using a set of guidewords in combination with the system parameters to seek meaningful deviations from the design intention. A meaningful deviation is one that is physically possible—for example, no flow, high pressure, or reverse reaction. Deviations such as no temperature or reverse viscosity have no sensible, physical meaning and are not considered. The team concentrates

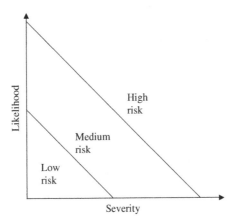

*Figure 1.1 Illustrative risk graph.*

on those deviations that could lead to potential hazards to safety, health, or the environment. It is important to distinguish between the terms hazard and risk. They have been defined[8] as follows: a "hazard" is a physical situation with the potential for human injury, damage to property, damage to the environment, or a combination of these. A "risk" is the likelihood of a specified undesirable event occurring within a specified period or in specified circumstances. It can also be expressed as a combination of likelihood and severity, as illustrated in Figure 1.1.

In addition to the identification of hazards, it is common practice for the team to search for potential operating problems. These may concern security, human factors, quality, financial loss, or design defects.

Where causes of a deviation are found, the team evaluates the consequences using experience and judgment. If the existing safeguards are adjudged to be inadequate, then the team recommends an action for change or calls for further investigation of the problem. The consequences and related actions may be risk-ranked. The analysis is recorded and presented as a written report which is used in the implementation of the actions.

To avoid misunderstanding and confusion with other forms of process hazard study or project hazard review, the term HAZOP study is reserved for studies which include the essential features outlined earlier in this section. Descriptions such as "coarse scale HAZOP" or "checklist HAZOP" should be avoided.

# CHAPTER 2

# Process Hazard Studies

Within the process industries, significant attention has been given to the development of comprehensive safety management systems (SMSs) or SHE management systems with the objective of protecting workers, the public, and the environment. There are also requirements within legislation such as the Seveso II Directive, European Union (EU) Directive 96/82/EC,[9] and subsequent country-specific legislation, requiring those companies handling hazardous materials to have in place an adequate SMS and to fulfill specified obligations. These requirements range from the preparation of Major Accident Prevention Policies to submission of detailed safety reports to a competent authority. Other non-EU countries have similar legislation—for example, the Office of Safety and Health Administration (USA) (OSHA) regulation 29CFR, Part 1910.119 (1992), Process Safety Management (PSM)[10] in the USA. An integral element of such systems is the use of systematic techniques for the identification of hazards. In addition to meeting legal requirements, there are considerable business benefits to be gained from the use of a systematic and thorough approach to hazard identification. These benefits include improvement of quality, faster start-up, and a reduction of subsequent operability problems.

For a new project, the greatest benefit is obtained by carrying out a number of studies throughout the design process. One such sequence is the Hazard Study (HS) methodology developed by ICI which used six stages.[4,11] Each study verifies that the actions of previous studies have been carried out and signed off, and that the hazard and environmental issues have been identified and are being addressed in a timely and detailed manner.

## 2.1 HS 1—CONCEPT STAGE HAZARD REVIEW

In this first study, the basic hazards of the materials and the operation are identified and SHE criteria set. It identifies what information is needed and the program of studies required to ensure that all SHE issues are adequately addressed. The aspects covered may include

HAZOP: Guide to Best Practice. DOI: http://dx.doi.org/10.1016/B978-0-323-39460-4.00002-5

reaction kinetics, toxicity data, environmental impact, and any special process features that need further evaluation. In addition, any constraints due to relevant legislation are identified. A decision may be taken on which of the remaining hazard studies (two to six) should also be undertaken. It is also important at this early stage to apply the principles of inherent SHE[12,13] within the design. This aims to eliminate, avoid, or reduce potential hazards in the process.

## 2.2 HS 2—HAZID AT FRONT-END ENGINEERING DESIGN (FEED) OR PROJECT DEFINITION STAGE

This study typically covers hazard identification and risk assessment, operability and control features that must be built into the detailed design, and any special environmental features to be covered.

It is important that the safety integrity levels (SILs)[14,15] of any safety instrumented systems (SISs) are addressed during this study as the design will still be flexible and simple design changes may be applied which will reduce the SILs and so simplify the design.

At the end of HS 2, the level of development of design and piping and instrumentation diagrams (P&IDs) would be "approved for design" (AFD). All the main features should have been added but the finer ones will not. It is helpful to examine the AFD diagrams for the more blatant errors using a form of checklist. An example is given in the final section of this chapter.

## 2.3 HS 3—DETAILED DESIGN HAZARD STUDY

This normally involves a detailed review of a firm design aimed at the identification of hazard and operability problems. Relief and blow-down studies, area classification, personal protection, and manual handling may, if appropriate, be included at this stage. HAZOP studies are normally carried out at this stage.

## 2.4 HS 4—CONSTRUCTION/DESIGN VERIFICATION

This review is performed at the end of the construction stage. The hardware is checked to ensure it has been built as intended and that there are no violations of the designer's intent. It also confirms that the actions from the detailed design hazard study are incorporated, and operating and emergency procedures are checked.

## 2.5 HS 5—PRE-COMMISSIONING SAFETY REVIEW

This examines the preparedness of the operations group for start-up and typically covers training, the final operating procedures, preparation procedures, and readiness for start-up including function testing, cleanliness, and purging. Confirmation of compliance with company and legislative standards is done at this stage, for example, under the Pre Start-up Safety Review (PSSR) required under the OSHA PSM legislation in the USA.

## 2.6 HS 6—PROJECT CLOSE-OUT/POST START-UP REVIEW

This study, carried out a few months into the production phase, confirms that all outstanding issues from the previous five studies are complete and seeks any lessons that might give useful feedback to future design work.

In addition to these six studies, two more may be included. These are usually referred to as study zero and study seven, to fit with the numbering scheme used above.

## 2.7 HS 0—CONSIDERATION OF INHERENTLY SAFER OR LESS POLLUTING SYSTEMS

Study zero takes place between the Research and Technical Departments before the concept stage. It attempts to identify and to incorporate the inherently safer and greener ideas as early as possible so that they will be part of the final design.

## 2.8 HS 7—DEMOLITION/ABANDONMENT REVIEWS

This study can take place before the final shut down but the objective of the study is to identify those issues which should be dealt with during the demolition process. It should address issues such as cleaning methods and standards, size reduction, recovery and recycle of working inventories, recycle of equipment, safe disposal of nonrecyclable materials/equipment, and location of potentially harmful/toxic materials in the equipment or soil. In addition it should address the integrity of lifting devices/brackets, access routes, and the sequence of removal bearing in mind that some equipment may be supporting other equipment.

## 2.9 OVERVIEW OF HAZARD STUDIES

The relationship of HS 1–6 to the project life cycle is shown in Figure 2.1. Experience shows that the use of HS 1 and 2 ensures key conceptual issues are dealt with early in the life of the project and not left to the HAZOP study. Use of HS 1 and 2 makes the HAZOP study easier and faster.

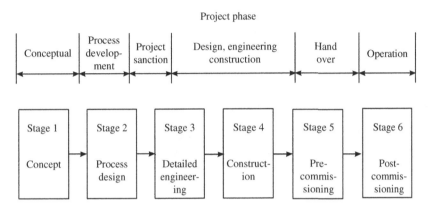

*Figure 2.1 Relationship of six-stage process study system to project life cycle.* (Source: EPSC, 1994, Safety Management Systems (IChemE, UK))

While HAZOP is one of the most flexible techniques for hazard identification, there are other identification and assessment techniques which can be used at the detailed design stage to supplement HAZOP (Figure 2.2).

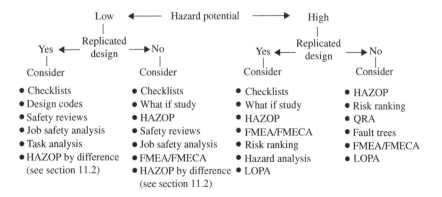

*Figure 2.2 Possible alternatives to replace or supplement HAZOP study as a detailed design hazard study.*

Whenever HAZOP study is the chosen technique, its use should be justified on the basis of complexity, inherent hazards, or the costs of the operation. While HAZOP study is ideally suited to novel processes, hazardous process or complex processes, it can equally be used in simple and repeat designs although there may be fewer benefits. Considerable benefits can also be found from its use for modifications or for change of use of plant.

More information on the techniques used for hazard identification and the way in which these relate to an overall process for risk assessment is given by Pitblado[16] and by Crawley and Tyler.[2]

## 2.10 ILLUSTRATIVE CHECKLIST FOR HS 2

An example of a checklist for use at conclusion of HS 2 to ensure common problems have been considered and are covered before the detailed design work begins.

- Are piping materials of construction in compliance with codes?
- Are spec breaks in the correct place? Are high- to low-pressure interfaces identified and given the correct treatment?
- Is the mass balance measured and achieved by the controls?
- Do any control parameters require independent verification?
- Do the protective systems (and SIS where appropriate) give adequate protection against the known hazards?
- Do SIS shut downs have pre-alarms?
- Have low flow conditions at start-up and shut down been addressed? Are any vents, recycle lines, and bypasses required?
- Are the standards of isolation appropriate for the risks?
- Is lagging appropriate to the piping codes?
- Are maintenance vents and drains shown?
- Are there traps in vessels that may need special features for entry?
- Are there any possible settling out points in piping or equipment which may need treatment?
- Has over pressure protection been applied as appropriate?
- Does the over/under pressure relief indicated look "adequate"?
- Is the design of the relief collection consistent? Can incompatible materials mix in the system?
- Do the vents go to a safe location?
- Do the drains go to a safe capture system?

- Is the design of the vent collection consistent? Can incompatible materials mix in the vent system?
- Is the design of the drain collection consistent? Can incompatible materials mix in the drain system?
- Are the main parameters (pressure, temperature, flow, and level) adequately controlled and are there adequate diagnostics to assess problems?
- Are line slopes shown and are they appropriate?
- Are there any pockets in the lines which may require drains?
- Are liquid capture bunds required anywhere for S or E requirements?

# CHAPTER 3

# The HAZOP Study Method

## 3.1 ESSENTIAL FEATURES

A HAZOP study is a structured and systematic examination of a planned or existing process or operation. At the outset of the study, the team creates a conceptual model (design representation) of the system or operation. This uses all available, relevant material such as a firm, detailed design, an outline of operating procedures, material data sheets, and the reports of earlier hazard studies. Hazards and potential operating problems are then sought by considering possible deviations from the design intention of the section or stage under review. The design intention is a word picture of what should be happening and should contain all of the key parameters that will be explored during the study. It should also include a statement of the intended operating range (envelope). This is usually more limiting than the physical design conditions. For those deviations where the team can suggest a cause, the consequences are estimated using the team's experience and existing safeguards are taken into account. Where the team considers the risk to be nontrivial or where an aspect requires further investigation, a formal record is generated to allow the problem to be followed up outside the meeting. The team then moves on with the analysis.

The validity of the analysis obviously depends upon having the right people in the team, the accuracy of the information used, and the quality of the design. It is normally assumed that the design work has been done in a competent manner so that operations within the design envelope are safe. Even where this is the case, the later stages of the project must also be carried out correctly—that is, engineering standards are followed and there are proper standards of construction, commissioning, operation, maintenance, and management. A good HAZOP study tries to take account of these aspects and of the changes that can reasonably be expected during the lifetime of the operation. A study will sometimes identify problems that are within the design limits as well as problems which develop as the plant ages or are caused by human error.

HAZOP: Guide to Best Practice. DOI: http://dx.doi.org/10.1016/B978-0-323-39460-4.00003-7

A key feature of timing of a HAZOP study is that the design must be firm and the P&IDs must be frozen—a situation that requires management commitment and forward planning.

## 3.2 THE PURPOSE

One purpose of a HAZOP study is to identify and evaluate any remaining hazards within a planned process or operation that were not identified or designed out in earlier stages. The hazards may be several types, including those to people and property, both on- and off-site. It is also important to consider the potential effects to the environment. Regardless of the type of hazard, many have directly related financial consequences.

HAZOP studies are also normally used to identify significant operability or quality problems and this will be included as a defined objective of a study. A survey of EPSC members carried out in 1999 as part of the preparation for the first edition of this Guide found that over 90% of the respondents included significant operability problems in the scope of the search. Operability problems arise through the reliability as well as the manner of the plant operation, with consequences such as downtime, damaged equipment, and the expense of lost, spoilt, or out-of-specification product leading to expensive re-run or disposal costs. The need to consider quality issues varies greatly with the details of the operation but in some industries it is a crucial area. Of course, many operability problems also lead to hazards, giving a dual reason for identifying and controlling them. A HAZOP study may also consider quality issues in the proposed design.

It is advisable to cover aspects of maintenance operations, including isolation, preparation, and removal for maintenance since these often create hazards as well as an operability problem. Where there are manual operations or activities, it may be necessary to analyze the ergonomics of the whole operation or activity in detail.

## 3.3 LIMITATIONS

Difficulties may be caused by inadequate terms of reference or poor definition of the study scope. The intention of a HAZOP study is not to become a re-design meeting. Nevertheless, some actions may result in changes to the design and potential problems may be found within the intended range of operation.

The analysis of problems within a HAZOP study is normally qualitative although, increasingly, simple risk assessment is used to help the team to decide on the need for action and the action itself. Some of the problems may need a fuller quantitative analysis, including quantitative risk assessment (QRA). This would be done outside the HAZOP meeting.

A HAZOP study is not an infallible method of identifying every possible hazard or operability problem that could arise during the actual operations. Expertise and experience within the team is crucial to the quality and completeness of a study. The accuracy and extent of the information available to the team, the scope of the study, and the manner of the study all influence its success. Only a systematic, creative, and imaginative examination can yield a high-quality report but even then, not every potential problem will necessarily be found. Additionally, the study will only be effective if the issues identified during the study are resolved and put into practice. Some important factors for success are listed in Chapter 12.

# CHAPTER 4

# The Detailed HAZOP Study Procedure

The actual study must proceed in a carefully planned, systematic manner to cover all of the selected aspects of the process or operation. It is normal to cover a continuous operation by dividing it into sections and working from an upstream starting point. A batch process or a procedure is divided into sequential steps and these are taken in a chronological order. The division of a process into sections or steps is described in more detail in Section 5.3 and illustrated in Appendices 3–5. The pattern of analysis for an individual section or step[a] is shown in Figure 4.1, and its main elements are described in the following sections.

## 4.1 THE DESCRIPTION AND DESIGN INTENTION

It is essential the team begins with a full understanding of the section or stage to be analyzed, either knowing the existing situation or having sufficient information to be able to form an adequate conceptual model. A full description should be developed, including all the key parameters, and the HAZOP report should include the design description.

Next, a design intention for the step is formulated and recorded. This should include a statement of the intended operational range (envelope) so that the team can recognize any situations lying outside this range as deviations. The design intention may be interlinked with the step description and hence to the design parameters of the equipment.

It is good practice to develop a comprehensive design intention, clearly linked to the drawings being used, which can be referred to during the search for deviations. A design intention may refer to equipment items in the section, to materials, conditions, sources, and destination, to changes or transfers, as well as to the means of control and timing of a step. It not only refers to plant equipment but covers what is intended to be done within the section being analyzed.

---

[a]Other terms sometimes used in place of section or step include node, stage, and part.

HAZOP: Guide to Best Practice. DOI: http://dx.doi.org/10.1016/B978-0-323-39460-4.00004-9

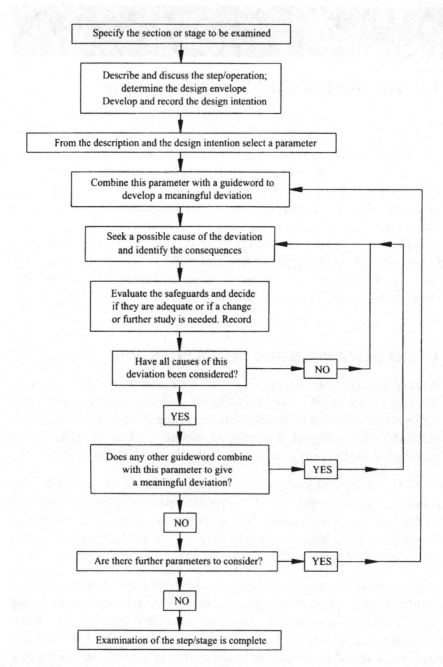

*Figure 4.1 Flow diagram for the HAZOP analysis of a section or stage of an operation—the parameter-first approach.*

The recording of the design intention should include sufficient information to enable a later user of the records to understand the picture developed and used by the HAZOP team during their study.

## 4.2 GENERATING A DEVIATION

The next step is to generate a meaningful deviation by coupling a guideword[b] and a parameter.[c] A deviation can be generated by taking a parameter and combining it with each guideword in turn to see if a meaningful deviation results (the parameter-first approach). This is the method described in Figure 4.1. The alternative approach is to take a guideword and try each parameter in turn (the guideword-first approach). More details of the guideword-first approach are given in Appendix 1, pages 95–97.

The standard set of guidewords for process plant is listed in Table 4.1, alongside their generic meanings. The first seven are normally used, with the others included if appropriate. As the purpose of the guidewords is to assist the team in a creative and thorough search for meaningful deviations, it is important to select a set that works well for the problem being studied. Variations of the standard set may be tried or others added to the list. Some companies have developed their own set of guidewords for particular technologies.

While clear recommendations can be made as to which guidewords should be considered, it is not possible to provide such firm advice regarding parameters. The selection of parameters is a task each team must address for each system studied. Table 4.2 gives examples of parameters that might be used in the analysis of a process operation. This list is not exhaustive but is intended to show the depth and breadth of the parameter and guideword search that can be used. It must be emphasized that many of the parameters listed will not apply to every issue or process as parameters relate to the individual system, process, or operation being studied.

[b]The term "guideword" is used here for an action word or phrase such as "no," "more of," and "as well as." Other authors have used alternative terms such as keyword.
[c]The term parameter is used here as the generic name for a variable, component, or activity referred to in the stage under study—for example, flow, pressure, transfer, and measure. Many alternative terms have been used, including keyword, property word, element, and characteristic. We discourage the use of keyword for either guideword or parameter as it may lead to confusion.

| Table 4.1 Standard guidewords and their generic meanings | |
| --- | --- |
| **Guideword** | **Meaning** |
| No (not, none) | None of the design intent is achieved |
| More (more of, higher) | Quantitative increase in a parameter |
| Less (less of, lower) | Quantitative decrease in a parameter |
| As well as (more than) | An additional activity occurs |
| Part of | Only some of the design intention is achieved |
| Reverse | Logical opposite of the design intention occurs |
| Other than (other) | Complete substitution—another activity takes place OR an unusual activity occurs or uncommon condition exists |
| Other useful guidewords include: | |
| Where else | Applicable for flows, transfers, sources, and destinations |
| Before/after | The step (or some part of it) is effected out of sequence |
| Early/late | The timing is different from the intention |
| Faster/slower | The step is done/not done with the right timing |
| *Interpretations of the guidewords for computer-controlled systems (programmable electronic system, PES) are given in the IEC HAZOP Application Guide.[7]* | |

| Table 4.2 Examples of possible parameters for process operations | |
| --- | --- |
| • Flow | • Phase |
| • Pressure | • Speed |
| • Temperature | • Particle size |
| • Mixing | • Measure |
| • Stirring | • Control |
| • Transfer | • pH |
| • Level | • Sequence |
| • Viscosity | • Signal |
| • Reaction | • Start/stop |
| • Composition | • Operate |
| • Addition | • Maintain |
| • Monitoring | • Diagnostics |
| • Separation | • Services |
| • Time | • Communication |
| • Aging | |

The extent of this list emphasizes the need for the team to form a clear conceptual model of the step and to use it to decide which parameters should be used in the search for possible deviations. When seeking deviations it must be remembered that not every guideword

| Table 4.3 Examples of meaningful combinations of parameters and guidewords | |
|---|---|
| Parameter | Guidewords That Can Give a Meaningful Combination |
| Flow | None; more of; less of; reverse; elsewhere; as well as |
| Temperature | Higher; lower |
| Pressure | Higher; lower; reverse |
| Level | Higher; lower; none |
| Mixing | Less; more; none |
| Reaction | Higher (rate of); lower (rate of); none; reverse; as well as/other than; part of |
| Phase | Other; reverse; as well as |
| Composition | Part of; as well as; other than |
| Communication | None; part of; more of; less of; other; as well as |

combines with a parameter to give a meaningful deviation. It is a waste of time to discuss combinations which do not have a physical meaning. Some examples of meaningful combinations are given in Table 4.3. Many parameters will emerge from the step description and statement of the design intention, provided it is explicit and comprehensive. In addition, a good team is likely to identify further parameters during the examination, particularly for the later guidewords "as well as," "part of," and "other than." It is good practice to apply all of the guidewords to the design intention before leaving a node.

Most of the combinations in Table 4.3 have obvious meanings but, as an example of the subtlety possible in HAZOP study, reverse pressure is included. It may apply to the situation in twin tubing where the pressure in the annulus between the outer and inner tube may be sufficient to crush the inner tube.

HAZOP study is most effective when it is a creative process, and the use of checklists for guidewords or parameters can stultify creativity. Nevertheless, checklists can be helpful for an experienced team. Illustrations are given in Appendix 2, pages 99–100.

## 4.3 IDENTIFYING CAUSES

Once a meaningful deviation has been identified, the team then seeks a cause. It is worth noting at once if the consequences are trivial as there is then no point in searching for causes. If there are likely to be several causes, as with the deviation "no flow" in a pipeline, it is very helpful to have a short brainstorming session to identify as many causes as possible, remembering that causes may be related to human factors as well as to hardware items. In seeking causes (and evaluating consequences), it is essential that all members of the team take a positive and critical, but not defensive, attitude. This is particularly important for any members responsible for the design. It can be useful to create and use a databank of frequently occurring causes to ensure no common causes are overlooked. If this is done, however, it should not be allowed to affect the creativity of the team or become the principal source of causes.

Although only realistic causes need to be discussed in detail, a judgment on this cannot be made without taking account of the nature and seriousness of the consequences. Acceptable risk involves an assessment of both frequency and severity so it is impractical to completely separate the discussion of cause and consequences in a HAZOP analysis. Sometimes this results in an action to assess the risk by more detailed analysis outside the HAZOP study meeting, for example, where a major consequence could occur as the result of a combination of causes. The term "realistic" implies a consideration of the likely frequency of a cause. If only minor consequences ensue, then even high-frequency causes may be ignored. In effect, a risk assessment is made based on a combination of the frequency of the event and the seriousness of the consequences. Experienced teams have little difficulty in this for most events. However, judgments as to the frequency at which causes are described as "realistic" are likely to differ from company to company and will certainly alter between countries due to different legislative approaches. In some circumstances, it may be best to analyze and record for even very low-frequency causes, perhaps with all the causes identified.

An alternative approach is to ignore the safeguards when evaluating consequences so that the ultimate effects are understood. Then each cause is considered in turn. Now the adequacy of the safeguards can be evaluated and the need for action determined.

It is important that causes are clearly described, as broadly similar causes may have distinctly different consequences. In these circumstances, it is necessary to distinguish and treat each cause separately. For example, pump failure due to a mechanical cause may cause loss of containment as well as loss of flow while pump failure due to an electrical cause may simply lead to loss of flow. So while it may sometimes be possible to group causes together, this should only be done where the team is sure that the consequences are identical for every cause.

Finally, before the discussion of a particular deviation is concluded, the team should consider all of the possible causes suggested.

## 4.4 EVALUATING CONSEQUENCES

The consequences of each cause must be carefully analyzed to see whether they take the system outside the intended range of operation. It is essential to fully identify all of the consequences, both immediate and delayed, and both inside and outside the section under analysis. It often helps to analyze how the consequences develop over a period of time, noting when alarms and trips operate and when and how the operators are alerted. This allows a realistic judgment on the likelihood and influence of operator intervention.

Where an effect occurs outside the section or stage being analyzed, the team leader must decide whether to include the consequences in the immediate analysis or to note the potential problem and defer the analysis to a later, more suitable point, in the overall HAZOP study. Whichever approach is adopted it is important that consequences outside the study section are fully covered, however distant they may be.

## 4.5 SAFEGUARDS (PROTECTION)

There are variations in practice as to when the existing safeguards and protection are noted and taken into account. One approach is first to analyze the outcome ignoring the existence of any safeguards such as an alarm, trip, or vent. Then, when the worst outcome has been identified, the safeguards are noted and the team moves to considering the need for action. This approach has the advantage that the team is alerted to possible serious consequences and misjudgments of the need for protection are less likely. Against this, it can be argued that it is unrealistic to ignore the in-built safeguards of a well-designed operation. Whichever approach is adopted, it is good practice to make note of the safeguards in the detailed records of the study.

## 4.6 RISK ASSESSMENT

Originally, little or no risk assessment was done in a HAZOP study, its purpose being the identification of hazard and operability problems. This is still a valid approach. However, if risk assessment is to be done during the study, the team needs an agreed approach covering:

- whether all problems will be assessed or only the high-severity ones;
- how it will be done;
- when it will be done.

It can be very time-consuming to do a risk assessment for every problem. However, if the team has a familiar, well-constructed risk matrix which is appropriate to that particular industry, they will become efficient at assigning likelihood and severity categories. A good software package helps by providing an easily viewed reminder of the matrix and may also allow different risks to be recorded for different categories of consequences such as environmental, process, or personnel injury.

The estimations of likelihood and severity are normally qualitative, typically in order of magnitude bands. They rely on the team's experience and judgment of similar events and will be uncertain, perhaps by as much as a factor of 3 (i.e., about one-half of an order of magnitude). A good team will quickly estimate frequencies as low as once in 10 years for common events. For lower frequencies, it may be necessary to make some analysis of the conditions needed for the event to occur and to do a rough quantification to get to lower frequencies. Inevitably, the uncertainty in the estimate will be greater for very low frequencies. When events of very low frequency, of one in 100 years or less, need to be considered, it is better to refer the problem to outside analysis by QRA or full Hazard Analysis and not to lose focus on the identification exercise.

The assessment is probably best made either after the team has clarified the consequences or following the discussion of the safeguards. Some companies choose to assess the risk at three stages:

1. unmitigated;
2. after safeguards;
3. after actions.

The advantage of this approach is that it shows the worst case consequences, the extent to which these are alleviated by existing safeguards, and then the effects of the proposed actions. This sequence makes it very clear how serious the problem is, the reliance on existing safeguards, and hence the need to ensure these are maintained during operations and the benefit, and hence the justification, for the proposed actions.

A further benefit of risk assessment after the consideration of the consequences is that minor problems are apparent and further discussion can be terminated.

## 4.7 RECOMMENDATIONS/ACTIONS

Several different approaches are in common use:

- After a potential problem is identified, it is always referred for investigation outside of the HAZOP meeting.
- At the other extreme, the team attempts, whenever possible, to deal with the problem and record a recommended solution to that problem whether engineering or procedural.
- The norm is for an intermediate approach where the team recommends a solution to the problem only if there is a breach of standards or if the team has unanimously agreed a solution which is within their authority to make. All other problems, particularly if there is no unanimity, are referred for further investigation outside the HAZOP meeting. This approach has the benefit that agreed hardware changes can be immediately marked on the working drawing and taken into account during the remainder of the study.

The approach used should be agreed in the definition of the study. Whichever approach is adopted, it is important that there is consensus among the team on any positive action, as well as on the causes and consequences. Also further causes, consequences, and deviations that might be associated with a change should be considered and covered within the HAZOP study. It is essential that all recommendations/ actions are unambiguous and clearly recorded so that they can be understood at a later stage in the project by non-team members.

Actions may be either specific or generic. The former is more common but, where a change might apply at several points within the design, it is simpler to make a generic action, so avoiding repetition and the possibility of different actions for similar problems in different parts of the process.

It is good practice to have an entry in the action column for every deviation and cause discussed, even if the entry simply states that no action is required because the existing safeguards are considered adequate, to show that the team concluded their discussion.

## 4.8 RECORDING

The conclusions reached by the team must be fully recorded, and it should be remembered that the HAZOP report typically represents the only comprehensive record of the study and of the operating strategy intended by the designers of the plant. The report should be regarded as one of the suite of key documentation handed forward to the operators of the project.

The selection of items to be included in the record are agreed during the planning of the study. It is important that sufficient detail is recorded for the potential problem to be understood outside the meeting by persons who were not present. The details of recording are discussed in Chapter 7. During the examination process, the team members should be aware of the details of the current record, either by it being displayed or by the leader stating what is to be recorded. In addition, team members should have an early opportunity to check the first draft of the meeting records.

## 4.9 CONTINUING AND COMPLETING THE ANALYSIS

In the parameter-first approach, the normal sequence is to consider in turn all causes of a particular deviation. When that is complete the same parameter is considered with another guideword to see if a meaningful deviation can be generated. This continues until all the guidewords have been tried. In practice a team quickly recognizes which guidewords to consider with each parameter. When all meaningful deviations have been examined, the team moves on to another parameter and considers this with all appropriate guidewords. The HAZOP analysis of the section or step is complete when the team can suggest no further parameters.

To get the best results from a HAZOP study, it is essential that the group functions as a team throughout, with every member feeling free to contribute and actually doing so. It is expected that a consensus will be reached at every stage of the analysis. If any team member is not satisfied with a conclusion or recommendation, then the team should aim to resolve the issue before moving on or turn it into an action for further discussion outside the meeting.

## 4.10 AN ILLUSTRATION OF THE HAZOP STUDY PROCESS

This simple example shows how a HAZOP study works. It is applied to a familiar task. The early stages are set out in full but the analysis is not completed, only going far enough to show at least one line of analysis for each guideword. You can easily add some more yourself.

Consider filling the fuel tank of a diesel-engined car as part of the operation of a new filling station. Assume the design of the filling station is complete and that it has been subjected to a full set of Hazard Studies. The intention here is to look at one function of the design. Consider a car driver arriving to take on fuel. Having selected this filling station, we consider what the driver has to do. A minimum set of steps is:

1. Select a filling bay that is not occupied.
2. Park so the filling hose can reach the inlet to the car's fuel tank.
3. Remove the cap from the fuel tank.
4. Determine which fuel is required—95-octane lead-free petrol, diesel, high-octane petrol, etc.
5. Place the fuel nozzle into the car's fuel tank inlet.
6. Start the flow of fuel.
7. Monitor the flow, stopping it when enough has been added.
8. Replace the fuel nozzle on the pump stand.
9. Replace the cap on the car fuel tank.
10. Pay for the fuel taken.
11. Drive away.

These could be made more precise but initial drafts of operating instructions rarely cover all situations.

Information must be collected for the study. This should include:

- The layout of the filling station showing entry and exit lanes, the number, position, and spacing of the pumps, and related buildings (the shop and pay point, tanker supply area and filling connections, the car wash, the compressed air and water supply station, etc.). Drawings and photographs of equipment items are required.
- The details of each typical pump station (if there is more than one style) with information on the number of fuel types available, the control system to be used, the display, and the flowrates. Drawings,

specifications, and photographs are the minimum; a team visit to the site would be useful. Normally a P&ID would be included.

- fuel properties;
- site drainage details and plans;
- fire safety measures and firefighting equipment;
- details on typical usage—fractional occupation of the available pump spaces, time per visit, range of amounts transferred, other traffic to and from the site (e.g., visits for shop purchases only);
- number of operators on-site and their general duties;
- frequency of supply tank filling and any restrictions placed on customer access during resupply;
- typical nonavailability of pumps, for example, due to shortage of fuel or individual pump failure;
- history of filling station incidents (specific to the operating company and in general).

We will assume that an experienced HAZOP study leader has been appointed to lead this study. The leader will review this information for general suitability and coverage and then think about the division of the steps of the operation (1–11 above) into stages for the study. The initial suggestion might be:

| Stage 1 | Steps 1–2 | Arrival and preparing for transfer |
| Stage 2 | Steps 3–9 | Filling the tank |
| Stage 3 | Step 10 | Paying |
| Stage 4 | Step 11 | Leaving |

We will look here at stage 2.

The team leader will need to assemble a suitable team. This might be:

- team leader (TL);
- site architect (SA);
- member of the site management (SM);
- representative from the pump manufacturers (PM);
- local operator (LO);
- representative user (RU);
- petrol company health and safety adviser (HS);
- scribe (TS).

After familiarization with the study data the team would discuss what is involved in stage 2 and draw up a design intention. This could be:

*To transfer diesel fuel from the selected fixed pump into the fuel tank of the car at the fastest rate compatible with safety. The amount transferred may be a chosen volume, a chosen value or the full tank capacity. The transfer will be controlled by an untrained member of the public and may be terminated manually or by automatic cut-off when the tank is full.*

The team, on advice from the leader, is using the standard set of seven guidewords, namely:

- No;
- More;
- Less;
- As well as;
- Part of;
- Reverse;
- Other than;

plus the additional ones of

- Where else;
- Before/after.

An initial consideration by the team of possible parameters gave the following ones (which may be extended by ideas suggested during the study itself):

- composition;
- flow rate;
- quantity;
- temperature;
- safety;
- control.

The following section gives examples of the team discussion, the first relating to high fuel flow:

TL    "I would like to discuss high flow of fuel into the car's fuel tank a little more. What are the implications of the failure of the *Dead Mans' Handle* on the filler or the failure to shut off in the case of high level in the car fuel tank?"

LO    "The fuel will spill out of the tank in an uncontrolled manner and go into the drain system where it will be caught in the interceptor."

RU    "Do we know if there is any level measurement or warning of overload of the interceptor?"

HS    "I think that there is."

TS    "I am making a note of the action on HS to verify this."

LO    "This raises some issues about the emptying of the interceptor both 'how?' and 'how often?'.'"

TS    "I am making an action on this between the LO and the HS."

TL    "Are there any more consequences associated with these causes of high flow?"

At a later stage, the team has a short brainstorming session to start the guide word "other/other than."

TL    "I suggest we start the use of the guideword 'other/other than' by brain-storming for possible deviations. Any ideas?"

LO    "A non-standard fuel container is filled."

PM    "Perhaps using a different fuel."

RU    "A car jacking is attempted."

HS    "Safety—an engine fire."

TS    "Car has a trailer or caravan attached."

RU    "Car won't restart or a puncture is noticed."

LO    "Leak of coolant, engine oil or other fluid from the vehicle."

SA    "Driver taken ill or appears so (drink, drugs)."

SM    "Extreme weather conditions—wind, frost, lightning, snow."

Table 4.4 shows extracts from the report.

Table 4.4 This is a selection from the report that could result from the study. Enough has been included here to illustrate each of the main guidewords at least once. An action placed on two team members means that they are both expected to be involved in resolving the problem. However, the responsibility for the response is placed upon the first named member

| Ref. | Deviation | Cause | Consequence | Safeguards | Action | On |
|---|---|---|---|---|---|---|
| 1 | No flow | Wrong initiating sequence used by the customer | Delay. Possible damage from wrong sequence. Sale may be lost. | Required sequence is usual for the UK and uses illuminated buttons on the pump panel. The site operator can select and speak to each station. | A1: Consider installing an alert to the operator whenever delay between removing hose and start of pumping exceeds selected time (say 20 s). | PM |
| 2 | No flow | Supply tank at low cutoff level | Delay and frustration for customer as cause not apparent. | Alarm to site operator of impending loss of supply. Operating procedure to cone off pumps with prepared signage. | A2: Review restocking arrangements against the expected demands to minimize this situation. | SM |
| | | | | | A3: Review operator training and testing. | HS |

*(Continued)*

| Ref. | Deviation | Cause | Consequence | Safeguards | Action | On |
|------|-----------|-------|-------------|------------|--------|-----|
| **Table 4.4 (Continued)** | | | | | | |
| 8 | **More** quantity | Customer error | Customer cannot pay; delay at till and at pump. | None | A5: Cover in training procedures. | SM |
| 9 | **More (high)** fuel flow | Dead man's handle on pump fails or the flow fails to shut off on high level in tank | Fuel spillage over side of car, onto ground, and into drain system. Possible fire. | Maintenance of the pumps. Interceptor within the drains. | A6: Check on the recommended maintenance procedures. | PM and SM |
| | | | | | A7: Check for level indicator and warning of interceptor overload. | HS |
| | | | | | A8: Review location and effectiveness of the first aid firefighting facilities. | HS |
| 10 | **More** time | Driver leaves car unattended (e.g., to shop in main store) | Pump blocked to other users. Uncertainty over "abandoned" vehicle. | None | A9: Establish procedure to deal with "abandoned" vehicles including emergency evacuation of the area. | HS |
| 13 | **Less** quantity | Low level in main supply tank | Customers cannot get fuel. | Low level warning on main supply tank. | A11: Check that resupply arrangements cover all likely rates of sale. See also A2. | SM |
| | | | | Operator training to cone off the affected pumps. | A12: Review | SM |
| 16 | **Reverse** entry of car into the pump lanes | Driver mistake or deliberate short cut taken | Confusion among other users and increased likelihood of on-site collision. | Signage | A13: Review the position and instructions on signs. | SA and SM |
| | | | | | A14: Consider if routing of entry/exit slip lanes can reduce occurrence. | SA |
| 18 | **As well as—** customer uses mobile phone | Customer ignores warnings | Possible ignition source—not likely with diesel but could be with petrol. | Warning notice at every pump station. | A15: Check on reality of the rumors of fuel ignition from mobile phones. | HS |
| | | | | | A16: Consider requiring till operator to warn phone users over built in speakers on pump station. | SM |

(*Continued*)

### Table 4.4 (Continued)

| Ref. | Deviation | Cause | Consequence | Safeguards | Action | On |
|------|-----------|-------|-------------|------------|--------|-----|
| 19 | Only **part of** sequence completed | Customer does not properly replace fuel nozzle on its stand | Transfer pump continues to run against closed valve. Payment cannot be made and customer must return to the pump. | Till operator can notify customer using the pump speaker but is unlikely to spot the problem before customer leaves the pump. | A17: Check with manufacturer how likely this is with the chosen design and what the alternatives are. | SM and PM |
| 21 | **Other** fuel container filled | Customer uses fuel can (perhaps as well as filling car fuel tank) | Pump is stopped and then restarted. May attempt to pump a different fuel. High-level cutoff may not work if container has a wide neck. | Not possible to pump separate fuel until payment made for first and pump zeroed. Restart with same fuel is possible provided nozzle not replaced first. | A18: Decide whether a timed cutoff should be included so restart is not possible after a selected time. | SM |
| | | | | | A19: Check whether high-level cutoff works in wide necked containers. | PM |
| 22 | **Other** event—carjacking attempted | Planned criminal activity | Risk of violence with injury (or death). | Warning to customers to remove car keys and not to leave car unlocked. | A20: Check wording and prominence of notices. | HS and LO |
| | | | Bad publicity inevitable. | | A21: Put up clear notice that CCTV is in use as a deterrent. | SM |
| | | | | | A22: Review emergency procedures to ensure this eventuality is covered and that training is provided. | SM and LO |

# CHAPTER 5

# Organizing a HAZOP Study

The detailed HAZOP study procedure described in the previous chapter needs to be fitted into an overall scheme. This chapter covers the preliminary organization while Chapter 6 covers the study itself.

## 5.1 DEFINING THE SCOPE AND OBJECTIVES OF THE STUDY AND ESTABLISHING THE BOUNDARIES

These two aspects are not independent of each other although they are discussed separately. This part of the planning contributes significantly to the eventual quality of a HAZOP study. A clear definition of the scope and objectives is crucial to the whole study and should define the responsibilities and authority of the team (see also Section 6.1). It should also cover responsibility for initiation of the study, for follow-up and implementation of the actions, and how any rejected actions are dealt with. A well-defined set of objectives helps prevent the team straying into areas that are not relevant to the study. Quality assurance (QA) and auditing of the study should also be planned from the outset so that the study and the recording facilitate any later checks.[17,18]

It is normal practice to include the identification of SHE risks as a principal objective, including risks to persons both on- and off-site, risks to property and plant, and all types of environmental risks. It must be made clear whether a further objective is to search for potential operability problems that by themselves have no SHE implications. These include reliability and maintenance issues, product quality or loss, and factors affecting plant life and productivity. Meeting regulations, company standards, and contractual performance requirements may also be a defined objective. Consideration should also be given to the intended uses for the study reports, in particular the need for audit and demonstration that a high standard has been reached in the identification of hazards and the assessment of risk. This may be required by both outside regulators and by insurance assessors. If it is expected that QA is required, the issues must be addressed during the definition of the scope.[17]

HAZOP: Guide to Best Practice. DOI: http://dx.doi.org/10.1016/B978-0-323-39460-4.00005-0

The boundaries for the study must be clearly defined as part of the specification of the study, including how any problems that extend beyond these boundaries will be handled. The first step is to identify those sections of the physical plant which are to be included. When boundaries are drawn, consideration is given to factors such as the nature of the process and the inherent hazards, the novelty of the operation, the complexity of the control system, and the relationship to other operating units. Where an existing plant is being considered, perhaps with a new process, it may be possible to omit from the study those sections that are used in a standard manner (although it may still be necessary to consider the effect that changes in the new section can have on the "standard" section). Also, if a standard unit or a proprietary unit is to be studied, it may be more effective to concentrate on the interfaces between that unit and other plant and operations. The unit may then be covered by comparison with a previously completed, full HAZOP study of the unit—an approach described as HAZOP by-difference. However, this should be done with care as there may be some small differences which could have a significant impact on operability.

When the study boundaries are drawn, there are likely to be interfaces with off-section elements of the plant such as drains, recycle lines, vents, and effluent treatment and perhaps sections of plant intended for occasional use for special purposes such as start-up lines. Consideration must be given to the inclusion of these within the HAZOP study, recognizing that interfaces with systems handling process materials are of potential importance. It is essential that such elements of the system are not overlooked simply because they are not on the drawings being used. The team needs to have an overview of the whole system so that important off-section causes and consequences are not overlooked, especially when associated with unusual events such as a plant trip. An interface review may be needed to clearly identify what crosses the boundaries.

There can be difficulties in setting boundaries when a modification to an existing plant is to be studied. As well as covering the modified sections of plant, it is usually necessary to extend the boundaries on either side to ensure that related sections, where causes of deviations that affect the modification and where consequences of deviations appear, are covered by the HAZOP study. It is a matter of judgment for the team leader.

Another aspect to be established at the outset of a study is the range of operational modes covered. For a continuously operating plant the main condition is the steady-state operation, but it is necessary to cover other modes such as start-up and shutdown (see Section 11.8), high and low rate running, hold conditions—particularly those anticipated during start-up—and changeover from one state to another. These may involve quite different pressure, temperature, or flow conditions compared to steady-state operation and may need detailed examination. Since many of them are discrete operations, they need to be examined using the batch HAZOP approach. For a batch operation, the additional modes could be special batches at the start and conclusion of a campaign or alternative control modes, such as fully automatic and partially manual. It is useful to think through the life cycle of the operation when deciding which alternative conditions should be included. It must be decided whether these alternative operational modes need separate examination or whether they can be covered under the guideword "other."

It is not possible to estimate the length of a study or to plan in detail until these boundaries have been established.

## 5.2 APPOINTING A TEAM LEADER AND SELECTING THE TEAM

### 5.2.1 The Team Leader/Facilitator

The selection of a team leader should be made at an early stage of the planning for a HAZOP study. An essential role of the team leader is to ensure the HAZOP methodology is used effectively and productively and so the leader needs to have a deep understanding and considerable experience of HAZOP studies. HAZOP study teams work best when there is clear leadership from an experienced leader. Desirable attributes for the leader include:

- wide experience of all the stages of process hazard studies, including QRA;
- extensive experience as a HAZOP study member and, preferably, some as a HAZOP study scribe;
- training in the leadership of HAZOP studies;
- technical competence and the ability to quickly understand the system and its operation;
- meticulous attention to relevant detail;
- good analytical thinking;

- motivational skills including the encouragement of creativity and open speaking;
- independence from the project itself with no other direct responsibility to the project manager other than completing the scope of the HAZOP study. The leader should be able to concentrate on the application of the method and the working of the team.

Independence from the project allows the leader to stand back and be able to take an objective, unbiased view. The leader acts, and is recognized to act, as an impartial person within the team. Freedom from other responsibilities is important because of the multiplicity of tasks already resting upon the leader. It is also advisable to check the suitability of the leader for the magnitude of the task, with only experienced and proven leaders given studies involving major hazards or other difficulties.

A good leader will have interpersonal skills that can be deployed to help the study group to function as a true team. These include the ability to listen, guide, and encourage individual contributions as and when required, to sense the unspoken feelings of individuals—perhaps indicated by their body language—to move the group toward a conclusion when a consensus is emerging and to work efficiently with the scribe. Within the HAZOP process, the leader directs the route taken through the stages, the parameters and guidewords. It may be necessary to vary from the preset plan by introducing new parameters and guidewords, to question doubtful issues and to defer examination—perhaps to gather more information—and to periodically review and remind the team of the key issues. The leader has responsibility for ensuring that the HAZOP study is functioning to a high standard and, if it is not, to call for changes which will restore these standards. In an extreme case, this could mean terminating a study until acceptable drawings, project information, or other team members are available.

A good leader working with a badly-selected team may be able to produce an adequate HAZOP study. The quality of the leader is critical and a good team cannot always negate the impact of a poor leader. If the team loses confidence in the leader, the study is doomed.

## 5.2.2 Scribe (Scribe/Recorder)

The scribe is another key individual in the team. The person chosen must be familiar with the HAZOP study method and usually has a technical background so that special explanations are not required. The scribe must be a good listener, always paying attention to details. In a major study, the scribe has no other role, as recording is a full-time task. In a small study—for example, a modification—or on a self-sufficient small site, the recording may be done by the team leader or a team member. Note that experience shows that the scribe cannot be contributing to a discussion and writing at the same time so this inevitably slows the study.

The scribe should establish a working relationship with the leader to become a helper, not just a recorder. Help can be given by noting suggestions and deferred problems and bringing them up at a later time, and by assisting in the selection of guidewords and parameters. More directly, the scribe starts to record as a consensus appears, without waiting for instruction or dictation from the leader. To do this the scribe must learn when to write if recording manually (or to type if recording electronically)—starting too soon is a waste of effort while waiting for a final version to be dictated can slow the progress of the study and interferes with the creative flow of the discussion. The team should know exactly what has been recorded. This may be done by the leader reviewing the entry, the scribe reading it back or, if a computer recording program is in use, having it projected so all of the team can see the record.

The choice of computer recording over a simple written record is a matter of preference as both methods can produce good records. If a projector is used—as is common practice—it must be understood that during the meeting the priority is to capture the essential points of the analysis; spelling is not a high priority as it can be improved later but there must be no ambiguity or incompleteness in the first record.

It is important the study records are produced consistently and speedily after each session, particularly in a lengthy study, and that final and agreed versions are produced.

### 5.2.3 Team Size and Membership

The size and composition of the team varies according to the complexity of the study. The minimum size for even the smallest study is probably four with some of the roles combined. Usually there is a core team of five or six. Occasionally, the number may be as large as eight or nine; team membership should be reviewed if this is exceeded. A typical team might involve the following personnel:

- team leader;
- scribe;
- process engineer;
- project/design engineer;
- plant/operations manager;
- operator (or commissioning team representative for a new design);

and one or more, as and when required, from;
- SHE expert (mandatory in some countries);
- research chemist;
- control/instrument engineer;
- maintenance/mechanical engineer;
- other specialists.

The essentials for the team members are that they have sufficient experience in the area of the operation and, as a group, have a comprehensive knowledge of the intended operation. In addition, the basic disciplines represented should ensure that a fundamental analysis of identified problems can be made. The key is the assembly of individuals who, together, provide the correct combination of basic disciplines and expertise, and general experience of the type of operation to cover all stages of the project. The disciplines needed may be a selection from chemical, mechanical, and electrical engineering, control systems engineering, chemistry, biochemistry, and other disciplines which offer specialist knowledge, when necessary, of reactors, piping, instrumentation, software, and metallurgy. General experience within the industry should average several years per member, including direct relevance to the intended operation with regard to the site, the method of control, and type of process. These latter aspects may be covered by the inclusion of a senior supervisor. Direct project involvement ranges from the initial research and development, earlier hazard studies, process design and the intended construction, commissioning, operations, control and maintenance groups.

It is possible that in complex processes which have a number of operating interests that the membership may be "dynamic." If this is the case the new team members must be given sufficient time to appreciate the direction and dynamics of the HAZOP study and be able to fit into the team at the appropriate time. Likewise a departing member should be satisfied that there is no further use for his/her team membership before departing from the team.

Compromises often have to be made when selecting the team. The ideal persons may not be available and there can be difficulties if a big team results from an attempt to meet all of the criteria suggested above. In a large team—say nine or more—separate conversations may start, making it difficult for the leader to capture everything and disrupting the unity of the team. The leader must make a balance to achieve the best possible results. The potential for conflict is greater in a large team; it is a responsibility of the leader to be alert to developing problems and to prevent them from becoming major issues.

Whenever a new design or a modification to an existing plant is studied, the team should always include a representative from the design group, from the operations team, and it is likely to include an instrumentation engineer. Other team members are then chosen according to the nature of the project. Possible further members could be a mechanical engineer, development chemist, electrical engineer, software specialist, and safety specialist. It may be possible for the safety specialist to also act as the team leader. A good team is likely to have several members who are closely involved with the project and one or more who have little or no previous connection. It should not consist solely of those who have worked full time on the design nor should it be totally independent of the project. The interaction between these two groups can improve the quality of the study. Some companies like to have an independent process engineer present to act as a critical questioner and challenger of assumptions, another role that can be filled by a truly independent team leader who has the necessary background.

In a normal HAZOP study where several meetings are required, it is important to have the core team members present at all the study meetings. Apart from the development of a team style and spirit, this maximizes the benefits from the growing understanding of the operation that comes as the study progresses. If one member of the team is from an operating team working on shifts, however, it may be impractical to have the same individual present at all the meetings. Specialists are added to the team when their expertise is needed for a particular section or stage.

It is important that some of the team have previous experience of HAZOP study; if some have not participated before, then training should be considered. Provided a few members are experienced, the others can be introduced by a short example or, if a single individual is involved, by a briefing and then by participation within the otherwise practiced team.

## 5.3 PREPARATION

It is important for the team leader to assemble and review the necessary data for the study and, if necessary, to reorganize the material for use by the team or to call for additional data. In this work, the leader may be assisted by a member of the design team and by the scribe. A site visit by the whole team can be very useful when an existing plant is studied but this is not always practical.

Knowledge of previous incidents on the plant or process being studied is important. A search should be made of appropriate accident databases to identify historic incidents relating to the type of process and to any specialized equipment being used. These may be corporate or international databases.

The boundaries for the study must be known and clearly stated. In the case of a continuous process, the boundaries will start at the feedstock and end at the final product distribution. It may be appropriate for a member of the plant producing the feedstock and another from the final user to partake in the appropriate phase of the HAZOP study. Minor changes in product specification as the result of a modification must be treated with care.

It is also necessary to plan the route the study will follow through the process; this must be done with care. There will be a number of branches and also heat exchangers which have process fluids on both sides. As a general guidance it is best to examine the route using the logical process flow but where there are branches or interactions, such as heat exchangers, it is appropriate to discuss the major issues that must be examined at the branch or on the other side of the heat exchange and then to make note of these for discussion at a later date. For example, the reduction in flow (low flow) on one side may have an impact on the other side of the heat exchanger and that process stream. Likewise a spared pump is a potential branch. It is appropriate to examine the interaction between the pumps but not to carry out a full HAZOP on each unit as that would be a repetition of effort. A note to the effect that the units are identical and only one plus the interactions was studied would be appropriate, provided the P&ID is truly identical in each case. Another issue might be the use of spared heat exchangers on fouling duty. The changeover and the impact of two heat sources on the process and pressure relief must be considered in detail. (See also the notes on vents and drains in Section 11.6.)

As a generalization a continuous process will start up (and shut down) in a "low flow mode." This must be part of the study. Likewise the start-up and shut down will involve dynamic changes in pressure and temperature (see Section 11.8). The HAZOP study must examine the whole envelope of these parameters. In some cases, particularly compressors and pumps, the low flow mode may require special start-up facilities which will only be used very occasionally. These must be examined with the same rigor as the rest of the process.

Following shut down the HAZOP study must examine the disposal or storage of off-specification materials which may contain contaminants including corrosive and toxic by-products.

This preplanning work is normally undertaken by the team leader, often with help from a member of the design team.

## 5.3.1 Continuous Processes

For a good HAZOP study to be possible, it is necessary for the team to have a complete and accurate process representation. For continuous operations this is based upon the P&ID, supplemented by design specifications and other details used in the design work, including flow diagrams, and material and energy balances. Further items are the intended operating conditions—usually expressed as a range—the operating and control philosophy and methods, and equipment and instrument specifications. Relief settings are essential to allow verification of pipe and vessel specifications, and alarm and trip settings should be normally available. In addition, there should be material properties, hazards and known operating problems together with the basis for safety, site details, and demands on the site services, and the results of earlier hazard studies including any material and reaction investigations. Information on vendor packages needs to be available.

Once the required level of detail has been obtained, the next planning task is the division of the P&IDs into sections and items to which the guidewords can be applied. Alternative terms may be used to describe the blocks selected for examination. The term "section" is used in this guide for part of a continuous process to which the guidewords are applied; another frequently used term is "node." The term "item" is sometimes used for vessels, exchangers, and so on, to which restricted or modified guidewords are applied. In a batch process, the terms step or stage are commonly used. A HAZOP study is a rigorous but time-consuming, and hence expensive, exercise. The higher the number of sections or steps, the more time-consuming is the study. Ideally, no sections should feature a process line junction and no steps should cover more than one element of the batch sequence. Experience suggests that some branching within a section and some step combination can be tolerated on systems other than those of very high hazard potential. Defining the intention helps demonstrate that the team have selected a worthwhile manageable node. Designation of the sections and steps is best undertaken by the leader at the outset of the study. Since this task is critical to the integrity of any HAZOP study, it is essential the leader has a proven track record in HAZOP management (see Section 5.2.1, page 31).

The selection of the sections is usually based upon the line diagram together with a description of the operating conditions. Often there are several possible ways of doing this division based upon the variation of parameters such as flow, pressure, temperature, and composition, at control points or junctions of lines with vessels. It is essential each section can be given a clear design intention and that a conceptual model can be constructed for the system in its given mode of operation. The size of the section contributes to effective analysis as there are problems in choosing either too small or too large a section. In the former case, there will probably be causes of deviations outside the section and consequences occurring upstream or downstream. There is a danger of these being overlooked. If too large a section is taken, then the design intention may be imprecise or very complicated so that it becomes difficult for the team to cover all possible meanings of each deviation. No simple, universal method can be given for the division into sections; experience is essential. The example in Appendix 3 provides an illustration.

It can be useful to mark each line and item of plant equipment as it is studied during the operating sequence. A final check can then be made to ensure all lines and items have been considered.

A further feature of continuous processes is that a high proportion of incidents and near misses occur when the plant is operated at conditions other than the steady state—for example, during start-up after maintenance, during commissioning, or shutdown. Therefore, it is important for the plan to give sufficient attention to these conditions and to the deviations that might occur in these situations—for example, too rapid heating or cooldown to the operating temperature, inadequate purging, or protective systems overridden. The team should also be mindful of the need for diagnostic instruments that will facilitate process performance monitoring and allow the assessment of local and wider heat and mass balances. These may be needed to allow assessment of heat exchanger performance, efficiency of pumps and compressors, and fouling of filters and piping, and can be of great importance during incidents.[19]

During start-up, it is sometimes necessary to override protective systems such as low temperature or pressure trips during what is an unsteady operation with a high risk of upset. The HAZOP study must pay particular attention to the process dynamics and risks with the protective system overridden and how the protective system will be reset (see also Section 4.5, page 19).

### 5.3.2 Batch Processes and Sequential Operations
The preparation work for a batch or sequential operation is usually more demanding than for a continuous process. In addition to the same range of background material, descriptions are also required of the detailed method of operation, the draft operating procedure, and the outline of the control sequence for a computer-controlled process. This time-dependent information is used in the division of the operation into stages for study, as illustrated in the example in Appendix 4. If a reaction is studied, the team needs information on the reaction process including heat and gas flows, exotherm onset temperatures, and the physical and chemical properties of the mixtures as well as the individual components. Again, the ability to define a design intention for each stage is an essential requirement.

One major difference between continuous and batch operations is that a physical item may need to be considered several times in a batch HAZOP study as the status, conditions, and design intention change. Thus, a reaction vessel may be considered during the addition of each component of the reaction mixture, during the reaction stage itself, and then during cooling and discharge. Another example is the common pipeline operation of pigging, where a typical sequence involving the loading of a scraper pig into a pig launcher at the start of the operation might be:

- Check the initial conditions.
- Open the LP flare valve and vent the vessel.
- Open the drain valve. Allow the launcher vessel to drain for 5 min. Then close the drain valve, leaving the launcher drained and at LP flare pressure.
- Close the LP flare valve and purge the pig launcher with inert gas by repeated pressurization to 5 barg and then venting to the LP flare until hydrocarbon gas tests show it is gas-free as measured at a representative sample point.
- Lock open the atmospheric vent valve and lock closed the drain valve and master isolation valves, leaving the vessel at atmospheric pressure.
- Check double block valve vents for any sign of leakage.
- Open closure door, load the scraper, and then shut the closure door. The scraper is now loaded but not launched.

The pig launcher is at the center of each of these steps but each one involves a different configuration and usually different contents and pressures. The main issues will differ from step to step as will the design intention. Hence, the parameters will vary between the steps. Each of these steps should be considered separately, using the full round of guidewords each time to ensure that the hazards are fully examined.

Further requirements for batch process HAZOP studies are intended trip settings for both the control system and the safety system, including those that take place between successive stages of the process. The HAZOP team must take into account the manner and timing of the resetting of the control and trip systems, covering both safety and operability aspects.

Another difference from continuous processes is the need for the team to picture the whole operation and the changing conditions throughout the plant at all stages of the batch. Without this understanding, interactions and consequences may be overlooked. Indeed, as the progress of one batch is followed through its various stages, a previous batch may still be in a connected section of the plant and preparations for a following batch may be underway. Thus, a deviation in the batch being followed through may cause an interaction with the previous or the following batch and so have consequences for these batches. The HAZOP study team needs to have a clear overall picture to enable it to identify such cross-connections; time-related flow charts, for example, a bar chart, are needed to give this picture. Such charts are unlikely to be available from the design work and, if needed, must be prepared for the HAZOP study. Good preparation and organization of the data can greatly enhance the prospects of success in batch operations.

### 5.3.3 HAZOP Study of a Procedure

Many operations, both simple and complex, are dynamic and are done by following a strict, predetermined procedure. Any failure to strictly adhere to the procedure or to ensure all the steps are done in the correct order may have the potential for a serious consequence. Start-up and shut down operations normally follow a sequence set out in a procedure, and these operations are known to be common occasions for problems and incidents. Hence, all procedures should be considered for evaluation by HAZOP study. The general principles of HAZOP study, as discussed in this monograph, hold good but the approach has to be altered to get the best results. This is described below and illustrated in Appendix 5.

The major difference from the examination of a steady-state operation is that procedures are carried out in a chronological order and so each step must be considered in turn and be challenged by appropriate guidewords. Thus, instead of traditional parameters such as *flow*, and guidewords such as *more*, potential problems are more likely to be uncovered by deviations such as *operation mistimed* or guidewords such as *too early/too late* or *out of sequence*. However, this does not preclude the use of the usual parameters and guidewords used in the HAZOP study of steady-state processes.

As ever, the team must have a clearly defined node and design intention for the step being examined. The node may be a piece of equipment such as a pump or a pig launcher. It may be a section of line, as might be selected for steady-state HAZOP study, but where the flow is not fixed but is dynamic due to moment-by-moment changes during the procedure. The node may include a number of simple operations that comprise a clear section of the procedure, for example, a series of valve changes in starting a pump or changes of flow, quantity, and temperature for the start-up, shut down, or conditioning of a reactor. Arriving at the best selection of nodes is often more difficult than when planning for a continuous process HAZOP study.

The choice of guidewords is also more challenging. The guidewords *too early/too late, out of sequence* or *incomplete*, could identify missed venting in the pump start-up or residues left in the reactor. Further possible guidewords are tabulated in Section 10.2 on Human Factors. Of course the standard guidewords given in Table 4.1 should also be used if they generate meaningful deviations with the parameters of the node.

Good practice, as with HAZOP study of steady-state operations, is to identify appropriate guidewords and parameters before the study is commenced, and it is the responsibility of the team to adapt the method of study to suit the problem being examined.

# CHAPTER 6

# Carrying Out a Study

## 6.1 PREMEETING WITH CLIENT

It is important that the HAZOP team knows what is required of it, and the Client (Project) has had an opportunity of discussing this with the team. This is particularly important when the Facilitator is an external consultant.

The starting point of the HAZOP should be a meeting between the Client and the Facilitator. During this meeting, there should be a discussion such that the Facilitator knows what the Client requires from the HAZOP and what can be delivered ("the Deliverable"). It is quite possible that each Client will have specific requirements. This meeting should prevent any possible misunderstanding which may affect the quality of the final deliverable. This meeting should draw up a Scope and Terms of Reference.

Consideration should be given to the following when defining the Scope of the study:

- The start and end points, which might be outside the immediate P&IDs (particularly for a modification or a plant in a sequence of many).
- That which *is included* and that which *is excluded* from the study. For example with a modification some operability issues may already be recognized by Operations. Should they be noted? Likewise should any "mothballed equipment" be inside or outside the scope bearing in mind that it could be brought back into operation without further analysis?
- The likely Parameter and Guideword Matrix to be used.
- Any particular issues that the Client feels might require special attention.
- The availability and competence of future team members.
- The spread of disciplines.
- In the case of a modification, the boundaries of the study both up- and downstream of that change.

HAZOP: Guide to Best Practice. DOI: http://dx.doi.org/10.1016/B978-0-323-39460-4.00006-2

- In the case of a modification, the Facilitator should satisfy him/herself that the P&IDs are of "Approved for HAZOP" quality. In the case of a modification, the Facilitator should satisfy him/herself that any interconnecting drawings (such as services) are both available and to the correct standard.
- The requirements or otherwise for risk ranking (outside the study) and against what criteria.

It is vital to note the status and revision of the information (e.g., procedure or drawings) to be used as the basis for the study as close out of actions normally requires them to be updated and significant changes may necessitate re-HAZOP.

Equally it is vital that team membership is limited to those who can contribute constructively and can contribute without external interference.

Consideration should be given to the following when defining the Terms of Reference:

- The method to be adopted (corporate or other).
- The recording format, spread sheet or other computer-based systems.
- In the case of a modification, how potential issues identified *outside* the immediate bounds of the study should be handled, recognizing that causes could be within the scope but that the consequences identified may be outside the scope of the system under review.
- The delivery date and reporting format. This must take account of how the actions will be passed on and tracked. It is good practice to ensure that the reason for any recommended action can be understood when that line of the HAZOP study record is read in isolation.
- The requirements for the follow-up HAZOP Study following resolution of the actions.

Finally, the Facilitator should have a view of the room to be used for the study and to satisfy him/herself that:

- the room is a suitable size allowing the tables to be arranged to the Facilitator's preference (e.g., horseshoe, round, or conference style);
- the heating/air conditioning, lighting, and ventilation are adequate;
- there is sufficient display space for large-scale drawings and adequate storage for supporting information such as line lists, pipe specs, and data sheets;

- there are sufficient power points and projection screen or a suitable wall area;
- there is communication from the meeting room but not necessarily to it. Wi-Fi links are useful to allow internet access to reference material;
- good refreshment and break facilities are to hand.

The facilitator may also wish to review the general support arrangements for the study such as the availability of the members and possible alternates, the travel and accommodation arrangements for those from off-site, and the possibility of changes to P&IDs during long studies.

## 6.2 PLANNING THE MEETINGS

The first requirement is a good estimate of the number of meetings required for the study. This can be made once the boundaries for the study are defined and the preliminary planning work, including the selection of the stages, has been done. The length of time needed to analyze a section or stage depends on its size, complexity, and the associated hazards. Experience from similar studies provides a good guide to the length of time to be allowed and hence the number of meetings needed.

An ideal arrangement is to have no more than three or four sessions a week, each limited to half a day. This is often impractical, however, and many studies are done as a full-time activity. In this situation, the leader must monitor the team's performance to ensure an acceptable standard is maintained. There should be an arrangement that allows all the team members to work from central documentation. Regular short breaks are advisable and interruptions to meetings should be prevented except for emergencies.

Team members should be provided with background details of the planned study in advance of the first meeting. This should include details of the scope and purpose of the study, essential design information, and an indication of the HAZOP approach to use, including a first list of guidewords and parameters. Normally, the team membership and details of planned meetings are included.

Full details of options and methods of recording are covered in Chapter 7. It is helpful to have a standard template for recording to ensure that none of the intended entries are overlooked. The style of recording should have been agreed at the premeeting (see Section 6.1), and the whole team should then be briefed.

## 6.3 THE STUDY MEETINGS

### 6.3.1 The Initial Meeting

In a long study, the first meeting differs from the bulk of the working meetings. After a reminder of the study objectives and scope, a brief overview of the project should be given, preferably by the individual most closely associated with the development work. This should cover the plant and intended operations as well as its relationship to the site, the services, and neighboring units. For a reaction system, the overview includes the process chemistry and basis for safe operations. It may be useful to give a short review of the HAZOP working method adopted, including discussion of the guidewords and parameters. This is particularly important if the Project is in any way out of the ordinary. Any queries about the precirculated material can be dealt with. This preliminary discussion aims to ensure the team has a common understanding of the project and it helps to establish the group as a functional team. It is also an opportunity for team members to learn of the skills and special competencies of each other.

If the team members have not worked together before, it can be helpful to discuss and agree a set of "ground rules." They may range from behavioral aspects such as listening to others, giving everyone a chance to speak, and not having simultaneous conversations to technical aspects such as how they will help the scribe and whether solutions will be sought to the problems identified. To be effective they need to be accepted ("owned") by the members although the leader, probably having most experience of HAZOP study, may suggest many. Working within the rules is best done by self-discipline of the members but the leader may, at times, have to remind the team of them.

### 6.3.2 The Detailed HAZOP Study Meetings

These follow an agreed plan, working as described in Chapter 4, concentrating on the identification and recording of potential problems for all SHE hazards and, if agreed, operating problems as well. It is recommended that the project's process engineer suggests a design intention for each section or step and specifies the safe design envelope for operation. The team can then discuss the design intention and refine it if necessary. The extent to which the problems are evaluated, ranked and solved, varies according to company policy and requirements. In the planning and definition of objectives for a study, it must be made absolutely clear what the responsibility and authority of the team is in these respects as failure to do so results in confusion and wasted time within the meetings.

There should be periodic reviews of the work, either at the completion of a section or stage or at the end of a meeting. As well as confirming the recorded details of the analysis, this encourages a check upon the progress of the study against the study plan. Reasons should be sought for any significant deviation. If ranking of frequencies or consequences is to be done, then the most efficient way is to do this is at the end of a session. Similarly, this is probably the best time for an outline discussion of SIL ratings if this is required. The whole team should receive a copy of the meeting records for checking as soon as possible after the end of a meeting in addition to any action notes assigned to them.

## 6.4 COORDINATING AND REVIEWING RESPONSES

The need to receive, check, and incorporate action responses arises as a study progresses. In a short study, this may be done at a special meeting after the whole study is complete. In a long study, it is normally done by periodically using part of a meeting to review the responses, after which they can be incorporated into the formal final record.

It is essential that when a recommendation for change is accepted, whether a hardware or a software change, the team agrees that it is an adequate solution to the original problem and also that it does not cause further problems by introducing new deviations. However, responsibility for the accuracy and adequacy of the response lies with the individual to whom the action was assigned, not the team. The responses generated from the action sheets should not be accepted unless they provide sufficient detail of their basis, including any necessary calculations, references, and justification for the proposal. These will become part of the HAZOP study records and hence will be part of any audit or review to show complicity with regulations.

Where a response generates significant design or operational change affecting the design intent then it may be appropriate to HAZOP study these as well.

## 6.5 COMPLETING AND SIGNING OFF THE REPORT(S)

The simplest and most common definition of completion for a HAZOP study is when all of the selected plant and operations have been examined and all of the problems identified during the examination have been considered. This includes collection of the responses and actions, and review and acceptance by the team (or by an authorized person or subgroup) as a satisfactory response to the identified problem. At this point a delegated person signs the detailed HAZOP study report as complete, that is, that responses have been received for all of the actions and that these have been reviewed and are considered to be satisfactory. It does not mean that all the recommendations have been carried out—that has to be covered under a separate management procedure. In major studies, the signing off may be staged. If a few responses remain outstanding, it is still possible to sign off the remainder provided company procedures include a secure follow-up mechanism. In this case, it is advisable that each outstanding action is allocated to a category—for example, to be done before start-up or may be completed during commissioning and so on.

## 6.6 FOLLOW-UP OF ACTIONS AND MANAGEMENT OF CHANGE

Responsibility for the implementation of actions, including any rejections, passes to the line management for the project, where an authorized person has the responsibility for signing off the actions as they are implemented. Although individual members of the team may have responsibilities for the implementation of some actions, the team as a group should not. If an action is rejected, the records should include the reasons for this, including a signed authority. There should be a formal check before the facility is commissioned to ensure that all actions identified during the study have been implemented or resolved. An established tracking system for actions is needed on major projects, and it may be done as part of a computer-based HAZOP recording package.

Once the HAZOP study is complete, it is important to introduce a system to minimize and control any further changes to the design. The implications of such change to the safety of the process must be considered by a structured Management of Change Procedure and, in some circumstances, it may be necessary to reconvene all or part of the HAZOP team.

The management of change must be a live system which captures and tracks all changes both before start-up and through the operational phase of the project.

It can also be beneficial to conduct a follow-up review, perhaps 12 months after the study, to draw out the "lessons learned." Beneficial changes can then be used to improve the Corporate Design Guides as in HS 2 (see Chapter 2).

# CHAPTER 7

# Recording and Auditing

The final records are all that exist to show the work done by the team. They may be revisited at a future date for a number of reasons, the most likely being as part of a Safety Case. All future uses of the records should be identified in order that the recording and reporting system is designed to meet these needs in an efficient way. The final reports should cover the why, how, when, and by whom of the study, as well as recording the necessary details of the analysis and actions of the team. They must be understandable by nonmembers of the team, even some years later.

The study team will use the records in a number of ways. Draft versions of the records are used by the team for checking after each meeting and generating action notes as well as for reference during the study. The final reports are needed for the implementation of the actions and to link to later process hazard studies. There may be regulatory and contractual obligations to be met and a requirement for audit, and they will be a key part of the plant safety dossier. They may represent the only detailed record of the intended operating strategy as developed by the design team. They may also be needed as a starting point for the HAZOP study of modifications. If a significant incident occurs during the lifetime of the operation, it is likely that regulatory authorities will wish to examine the records. Thus, the team should be mindful of the uses for which the final report may be required.

In addition, the records should provide an easy and clear understanding of the process and the equipment, be of especial use in the preparation of Operating Instructions as well as for troubleshooting and operator training and have the potential to improve the management of change (MOC). Hence, it is essential to anticipate the intended uses so that the HAZOP study file contains the necessary information, detail, and clarity to meet the requirements of each use. As a minimum, the marked up P&IDs, HAZOP tables, and all action responses are archived.

HAZOP: Guide to Best Practice. DOI: http://dx.doi.org/10.1016/B978-0-323-39460-4.00007-4

## 7.1 BACKGROUND INFORMATION

The amount of material in the final HAZOP study file depends upon the company practice for archiving of related materials. Items which are securely archived—for example, all the revisions of the P&IDs, the Material Safety Data Sheet (MSDS), and reaction hazard investigations—can be referenced without putting duplicates into the HAZOP study file. Otherwise all drawings, including the version used at the start of the study, and all other documentation are included or referenced. This includes any previous hazard studies referred to during the HAZOP study, site drawings, specially prepared material, outline operating or control procedures, and so on. Draft versions of the detailed meeting records need not be included provided the full final version is incorporated.

Additionally, a statement of the HAZOP study procedure should be given—for example, by reference to a company protocol. The name and a statement of company role should be included for each person in the team, including an attendance record. The remit of the study should be made clear and the hardware boundaries stated. The records should show the operational modes selected for examination and indicate how this was done—that is, by separate examination or by grouping a range of conditions within one examination.

## 7.2 SECTION HEADINGS

For each section or stage given a complete round of guideword examination, there is a header which explains the model used by the team. This identifies the section limits, its status and contents, and the means of control. There is also a full statement of the design intention, as developed by the team in their search for deviations. If the design intention is complex or extensive, then reference to an appropriate design brief or specification may be used.

## 7.3 THE RECORDING FORMAT FOR THE DETAILED EXAMINATION

The discussions of the team are normally recorded in a tabular format with a series of main columns, perhaps with some supplemental sections for each entry. The minimum set of columns is:

| Deviation | Cause | Consequence | Action |
| --- | --- | --- | --- |

However, it is commonplace to include an initial column for the "parameter," and after the "consequence" column it is also good practice to have one headed "protection" (or "safeguards"), particularly if these offer significant risk reduction. It is essential to have a numbering system, either numbering separately each entry or each action, and it is usual to link the action to the person responsible in an "action on" column. Thus, the norm is likely to be:

| Ref. | Parameter | Deviation | Cause | Consequence | Safeguards | No. | Action | Action On |
|------|-----------|-----------|-------|-------------|------------|-----|--------|-----------|

Further items to be considered are "response," "comment," "hazard category," "event frequency," and "event magnitude." From these last two it is possible to develop a risk ranking scheme. The choice of which items to include depends on the company style and the uses to be made of the records, as well as on the recording system used.

A risk matrix is needed if the event likelihood and severity classifications are to be used in risk ranking—indeed if this is to be done it is likely that the company will have an established risk assessment matrix. This may range from a simple three-by-three matrix to more complex methods. Not all users of HAZOP rank potential problems and, if it is to be done, a suitable matrix must be agreed before the study is started. As described in Section 4.6, three levels of risk may be shown: for the worst case, then after allowing for the existing safeguards, and finally after the recommended actions have been incorporated. The merit of this approach is that it explicitly shows the importance of maintaining the safeguards and of implementing the action. It must also be remembered that HAZOP studies are not well suited to the identification of high-consequence low-frequency events. These should have been identified in the earlier hazard studies.

## 7.4 THE LEVEL OF RECORDING

Once the format is determined, the level of detail of the recording is decided. Three levels are possible:

1. record by exception—only when an action results;
2. intermediate record—where an action results, where a hazard exists, or where a significant discussion takes place;
3. full record—all meaningful deviations are noted even if no realistic causes are found.

Recording by exception requires an entry only when the team makes a recommendation. This method serves the immediate needs of the study and provides a basis for implementation of the actions but is of little value in any subsequent uses. It is not recommended for general use. It may lead to shorter meetings and simpler reports, and be superficially attractive if there is pressure to complete the study within the project time constraints, but any economies are likely to be false ones as the fuller levels of recording have many later benefits.

At the intermediate level, a record is generated whenever there is any significant discussion by the team, including those occasions where there is no associated action. These include deviations identified by the team which, though realistic and unanticipated in the original design work, happen to be adequately protected by the existing safeguards. There are some important gains by recording at this level. One is the increased likelihood that these safeguards are maintained during the plant lifetime since their purpose is spelt out in the records. A second benefit is that the ground covered by the team is clear during an audit and to any later reader of the HAZOP study file; any deviation not recorded was either not considered a realistic combination of guideword and parameter or was thought to have no significant causes or consequences. A third benefit is the ease with which modifications can be analyzed by HAZOP study at a later date.

In full recording, an entry is included for every deviation considered by the team, even when no significant causes or consequences were found. At this level, each parameter is recorded with each guideword for which the combination is physically meaningful. It may even extend to listing the guidewords that were not considered as they did not give a meaningful deviation with the parameter examined. Reasons for recording in full are to conform to a company standard, or the high hazards involved, or to meet a legislative requirement, such as the OSHA legislation in the USA. Some shortening of the records may be possible by having standard entries to cover some common cases. For example, if no causes can be found by coupling a parameter with a group of guidewords, the term "remainder" can be written with the parameter and the phrase "no causes identified" put in the cause column or "no significant risk" in the consequence column. Full recording is obviously more expensive and leads to a very substantial file but does permit a full audit. It is therefore the preferred option for those industries that need to demonstrate the highest possible standard of safety management.

## 7.5 THE CONTENT

It is essential that all entries, whether causes, consequences, protection, or actions, are as clear as possible and identify unambiguously the items to which they refer, using vessel, equipment, and line numbers. If the records are too brief or otherwise inadequate, they may be open to misinterpretation so creating difficulties in the implementation of safety and operability into the final design.

## 7.6 COMPUTER RECORDING

Dedicated computer recording systems have been available for many years and are widely used, particularly for large HAZOP studies although standard word processing or spreadsheet programs are perfectly adequate. A list of commercial software has not been included here due to the problems of giving comprehensive coverage and maintaining an up-to-date listing. Gillett[4] gives some information while further sources are software houses and consultancy companies.

There are few disadvantages to using a computer recording system provided it is done well, and it is certainly worth considering if a handwritten record will later be word-processed. Probably the major disadvantage is that the package may force the recording, and perhaps even the HAZOP study, to be done in the way envisaged by the program designer. It is essential that the package allows the chosen style to be followed. It is also important to have a scribe who is familiar with the recording program and is able to type fast enough to avoid any delays to the meeting. Also, if the records are displayed on screen for the team to see, the display system must be powerful enough to avoid the need for a dimly-lit room. As with hand records, the forms can usually be customized to suit a company style. A great deal of preparative work can be done beforehand by the leader and the scribe and, even if changes are needed during the meeting, this is easily done.

There are some advantages with computer recording. During the study, the headers, which include the design intention, and earlier sheets are easily consulted and seen by all of the team. Single keystroke entries are made for parameters and guidewords and for frequently used phrases. If the team can view what is being recorded, then any disagreements or possible ambiguities are dealt with immediately. Databanks of possible causes, effects, and frequencies are held on the computer and consulted when needed. Draft records for checking are available for circulation by printing or email shortly after the end of a meeting and action notes can also be generated without delay. However, general circulation should only be done after the leader and scribe have checked them for spelling and meaning. Responses to action notes are easily incorporated into the records. Spell-checking facilities are normally available and it may be possible to search the program for individual words, names, or combinations—for example, to list all the individual records where responses are overdue. Some programs are used as a management tool for the study, and the more sophisticated programs are written for use with other process hazard studies. It is also worth noting that the recommendations/actions can be captured from the electronic records and quickly transferred into the MOC or HAZOP actions tracking system.

A different aspect of computer use in HAZOP studies is the expert system, designed to "conduct" a HAZOP study. A number of programs have been developed, but the present view is that a fully satisfactory system has yet to be written; indeed some think the creativity of a good team will never be duplicated by a computer. However, such programs do have some potential as a preliminary screening tool, for example, on P&IDs at the "approved for review" stage, since they can ensure no known cause or deviation is overlooked.

## 7.7 AUDITING A HAZOP STUDY

An audit or review of a completed HAZOP study may be done internally, to show conformity with company standards and to learn from the study, or externally. In this latter case, it is likely to be done by a regulator to confirm compliance with national codes or in the aftermath of an incident. All reviews will largely depend on the study records, and it is important this is recognized at the outset of the study. The list below covers some of the major points that should be examined in an internal audit. The key question is "did we do it properly?" whereas in an external audit it is "was it done properly?"

- General features
  - clear terms of reference and authority for the team;
  - scope of the study—start and end points; interfaces and links to facilities as a whole;
  - timing of the study within the project and facilities for the study team;
  - links to other project hazard studies;
  - the team
    - qualification of leader and scribe;
    - selection, competence, and experience of the team members;
    - continuity of attendance and use of specialists.
  - time available.
- Preparation and overview
  - P&IDs revision number and date. Should be either final design or as-built;
  - modes of operation selected for study (e.g., steady state, start-up, and shutdown);
  - other documentation made available (e.g., cause and effect diagrams, equipment specifications, isometrics, operating and control sequences, material hazard and data sheets, site plans, reports from earlier hazard studies);
  - any special preparative work done for a batch process or a procedure (e.g., batch progress charts, plant status chart);
  - node selection.
- Detailed report
  - style of recording which should include, as a minimum, clear reference number for each line, deviation, cause, consequence, safeguards, action, and action on;
  - marked off matrix of guidewords and parameters showing a comprehensive and imaginative use;
  - entries that can be precisely related to the P&IDs and are sufficiently detailed for the auditor to understand the meaning and outcome of the discussions;
  - a good design intention for each node, realistic causes, appropriate consequences, understanding of the design envelope and existing safeguards, etc.;
  - sufficient and appropriate depth in the search for causes. No excess of trivial items;
  - clarity when no causes are found for possible deviations or no action is required;

- significant perceived risks referred for more detailed risk assessment;
- archiving.
- Post study work
  - clear links to follow-up so all recommended actions can be traced to a final decision and implementation. This may require an audit of the formal closeout procedure;
  - evidence that the team has been able to review the outcomes of actions where further investigation was recommended.

# CHAPTER 8

# Training

## 8.1 TEAM MEMBERS

It is good practice for team members to have been trained in the HAZOP study technique. This is usually done through a formal training course covering the principles and the methodology of the technique. Experience can be given by the use of simple examples which allow the key points to be illustrated, as well as showing important aspects of preparing, leading, and recording the study. It is best if the course is quickly followed by experience in an actual study so that the learning points are reinforced and developed. Where an organization regularly uses HAZOP and has many experienced and practiced team members, it is possible to introduce a new member to a team after a brief introduction to the method. Subsequent learning is then done by working within a group of experienced practitioners. A recent development has been the creation of online courses for training team members. While these lack the interactions of group training methods, they can be useful for individuals with an urgent need for training or for those unable to access a group course.

## 8.2 SCRIBE

Formal training is not required provided the scribe understands the categories and level of recording needed. However, it is helpful if the scribe has been trained as a member and had sufficient exposure to HAZOP studies so that issues requiring to be recorded can be recognized promptly without instruction. If a computer recording program is chosen, it is essential the scribe is trained in its use so that the recording process does not delay the progress of the team. In a long study, it is helpful to have a scribe who is also a trained leader; this can allow roles to be rotated periodically.

HAZOP: Guide to Best Practice. DOI: http://dx.doi.org/10.1016/B978-0-323-39460-4.00008-6

## 8.3 TEAM LEADER

A team leader needs to develop and practice skills to:

• understand and use the HAZOP method and structure;
• manage the team effectively to optimize the contribution of all participants;
• use their own technical knowledge and experience lightly so as not to become the technical expert.

Thus, previous experience as a team member in a number of studies is important, preferably covering a range of different operations. Not all experienced HAZOP study members, however, can become successful leaders. It is also of value for the leader to have experience as a scribe in order to appreciate the particular problems with this task. Some formal training in leading, including practice undertaken alongside an established leader, is recommended. Initial experience should be gained on a low-risk process. As studies are led on higher risk processes, the degree of mentoring from a more experienced leader should be increased until the trainee becomes confident of how to address the nature of the hazards faced in such studies across a range of processes.

Obviously, the team leader must fully understand the HAZOP methodology and, during the meetings, must focus on applying this and making it effective. Equally important is man management—getting the best from the team members and the scribe. Having this skill is essential for good leadership; without it a leadership candidate should not be allowed to progress.

HAZOP study leaders should always critically review their performance after each study, if possible with the help of another experienced person from the team. It is important to maintain the skills of leading by regular usage, and this is helped if the company keeps a list of trained leaders with a record of the hazard study work done by each one.

More details[20] are available of the range of training methods that have been used.

# CHAPTER 9

# Company Procedures for HAZOP Study

If HAZOP studies are regularly used as part of a company's process hazard studies, a company procedure on the execution of a HAZOP study is recommended. This helps greatly in the uniformity of application and maintenance of standards and simplifies the reporting of any study carried out in accordance with the procedure. All of the relevant issues described in Chapters 2–8 of this guide can be included, as can those covered in the IEC guide.[7] It is important to give advice on the selection of suitable techniques for process hazard studies so that a HAZOP study is considered whenever it might be appropriate and is used when it is best suited.

A company procedure might cover:

- responsibility for initiating a study;
- where HAZOP might be applied and how it relates to other Hazard Studies;
- the purpose and the range and depth of study;
- any special requirements, for example, checking cause and effect diagrams;
- timing of the study, within the overall project and the detailed meetings;
- terms of reference and scope;
- detailed HAZOP methodology to be used;
- follow-up and implementation of actions;
- how records will be maintained;
- how the study integrates with the company's MOC procedure.

Further important aspects to be considered are:

- appointing a leader and scribe;
- team selection, composition, and training;
- documentation needed for the study;
- facilities for the team meetings;
- preliminary briefings;

HAZOP: Guide to Best Practice. DOI: http://dx.doi.org/10.1016/B978-0-323-39460-4.00009-8

- style and level of recording and the recording format;
- application (or not) of risk assessment with, if necessary, a suitable (corporate) risk matrix;
- any summary reports in addition to the main, full report;
- signing off of individual actions and the final report.

The HAZOP methodology could be expected to cover:

- approaches to continuous and batch processes;
- node selection;
- design intention;
- guidewords;
- parameters;
- identification of causes;
- consequence types (hazards, operability, financial, quality, reputation, etc.);
- consideration of safeguards;
- risk and SIL assessments;
- actions and recommendations—company preferences;
- human factor issues.

A company procedure must cover the general, basic HAZOP methodology, as indicated above, but it should also be tailored to the particular activities and types of process and materials used, as well as to the specific company arrangements for administration and organization. The standard guidewords should always be considered but there may be additional ones that have proved useful in previous studies. Likewise a list of parameters can be produced to act as a minimum set to be considered during a study. This can be particularly useful with the guideword "other/other than." There can be a list of specific issues that must be addressed, for example, in drilling where the clays can become agglomerated and then may choke the mud gutters. Another possible special issue is how shutdowns are addressed—there may be multiple levels of shutdown from the process shutdown (or "hold") to the ultimate shutdown for a major incident.

# CHAPTER *10*

# Advanced Aspects of HAZOP Study

## 10.1 HAZOP STUDY OF COMPUTER-CONTROLLED PROCESSES

The use of computers to control chemical processes in part or in entirety is now widespread, and many control devices contain some form of programmable logic. The number of reported incidents[21,22] in such systems demonstrates their need for effective hazard study. The introduction of computers has occurred since the original development of the HAZOP study method and so the technique has had to be adapted to cover such processes.

The reliability of PESs in safety-related applications is now covered by international standards such as IEC 61508[23] and 61511.[14] In the UK, the PES Guidelines[24] give detailed advice and point out the many types of problems that must be covered and give advice on how to deal with them. As well as anticipating random hardware failures, it is necessary to identify, and eliminate or minimize, systemic failures including those due to:

- errors or omissions in system specifications;
- errors in design, manufacture, installation, or operation of hardware;
- errors or omissions in software.

The specification and planning of a computer-controlled process is normally examined during HS 2; the detailed search for undetected errors and omissions occurs during HS 3.

Where safety-related systems are computer-controlled, they must be designed, installed, operated, and maintained to the specified standard for their purpose. The performance of the whole system must be demonstrable, including all elements—from the sensors, through the logic processors, to the actuators and final process hardware elements such as valves.

HAZOP study has proved to be an effective method of critical examination of computer-controlled systems. But first there must have

HAZOP: Guide to Best Practice. DOI: http://dx.doi.org/10.1016/B978-0-323-39460-4.00010-4

been an earlier HS 2 to identify the key issues and set out design requirements. Then the HAZOP methodology must be adapted to bring out the computer-related problems as well as the conventional hardware and human factor problems—the recommended approaches are described here.

### 10.1.1 Hazard Study 2

The title "Safety and Operability Review" (SOR) has commonly been used for the study carried out on a computer-controlled system at the time of HS 2. This should be done at an early stage in the project definition and design stages, as described in Chapter 2, in order to highlight the main hazards and to identify critical safety functions. It is carried out by a small team that includes at least one expert on computer systems.

The team uses a checklist or a set of questions that will help to highlight the key issues—it is not, therefore, following the traditional HAZOP method. Key topics to be considered include:

- safety critical protective functions;
- interactions between control and protection systems;
- the extent of hardwiring of controls, alarms, and trips;
- the degree of redundancy and/or diversity required;
- independence and common cause failures;
- input/output (I/O) arrangements for the control system;
- routing of data highways and their vulnerability;
- communication links and speeds;
- program storage method and the security;
- positioning and security of the computer hardware;
- likely consequences of system failure and of site power failure;
- construction of screen pages and alarm displays to assist troubleshooting.

Where significant hazards are present there will be a layered protection system starting with alarms allowing operator intervention, then actions by the control computer and, ultimately, hardwired or software initiated actions through SISs or safety-related protection systems and demands on the passive protection such as relief valves or actions driven by an independent computer.

## 10.1.2 Functional Safety and IEC 61508

IEC 61508[23] has implications for HAZOP study of computer-controlled processes. Its title has three key terms:

1. Functional safety—that part of the overall safety of the plant, process, or piece of equipment that depends on a system or equipment operating correctly in response to inputs.
2. The systems covered—any electric, electronic, or PES.
3. Safety-related—that is, systems that are required to perform a specific function to ensure risks are kept at an acceptable level.

The requirements placed on a safety-related system depend firstly on its function. This will be determined during Hazard Studies 2 and 3 when possible hazardous conditions are identified. Then the performance required for this safety function must be determined by risk assessment. A key part of the IEC 61508 methodology is the establishment of SILs for each hazardous condition, giving the level of reliability required of the protective systems. These can then be specified to match the SIL requirement, and individual components specified accordingly.

The standard uses four SILs. SIL1, the least demanding, specifies a probability of failure on demand (PFD) for low-demand events of between $10^{-1}$ and $10^{-2}$; the PFD range decreases by an order of magnitude for each step up in the SIL rating. To achieve SIL 3 or 4 is very demanding, usually requiring a high-integrity protective system with multiple sensors and actuators, voting systems, diversity, and redundancy. This is difficult to design, prove, and maintain and so most systems in the chemical process industries tend to be designed to operate with protective systems up to, but not beyond, SIL2.

In order to ensure effective implementation of a safety-related system, it is important to consider it throughout the life cycle of the project and process, including: scope, specification, validation, installation, commissioning, operation, maintenance, and modifications. IEC 61508 covers all these life cycle activities. This standard forms the basis for developing other standards for particular sectors of operations, including BS IEC 61511[14] for the process industries.

It is essential that these systems are defined early in the project so that the detailed design can satisfy the overall design intentions and to ensure that all computer-controlled safety-related functions are clearly recognized. If a positive decision is taken to operate with computer-controlled safety-related functions, then very high standards of design, proving, and maintenance will be required, and a Computer HAZOP (CHAZOP) study will almost certainly be necessary, to meet the reliability required by the Standards.

As well as meeting the required SIL, the systems should be designed to minimize demands on the protection systems from control failures, for example, by spurious trips. It is usual to address the requirements of the standard by having a combination of hardwired or passive protective devices while having the early warning alarms, initial actions, and duplicated safety actions passing through the control computer.

### 10.1.3 Enhanced HAZOP for Computer-Controlled Systems

The enhanced HAZOP extends the usual conventional HAZOP study to include aspects of the computer-controlled systems. It is suitable when the control system does not have safety-related functions and where the hazards are not exceptionally high and where there is a final level of hardwired protective devices.

The purpose of a conventional HAZOP is to consider possible deviations, using the full P&ID and other relevant design and operating information, and to single out for action those that have significant consequences not adequately controlled by the existing design. When a computer-controlled system is studied, the whole control loop from the field sensor through the logic solver and the final element must be considered. There will also be additional deviations that are due to random failures within the computer hardware. These could be the system as a whole, an individual input and output board, or the operator consoles.

Attention must also be given to the control sequences, and these must be defined in sufficient detail for the exact operating sequence to be understood for each section or stage being reviewed. In addition, it is important to know what status checks are being made by the computer and what actions will be taken by the control system if a fault is detected. The team needs to review and question these actions for each stage.

For an enhanced HAZOP, the information needed in addition to that normally used in a study of a conventionally controlled process includes:
- The specification of the computer hardware, the control language, and the programming method.
- The QA checks for the software.
- The signals from the plant instrumentation to the computer and the computer outputs to controls, valves, alarms, etc.

- The interface between the operators and the computer and the means of interaction allowed to the operators.
- Which safety features are hardwired.
- An outline of the control sequence on which the code will be based—probably as a logic diagram. Sufficient information is needed so that the intended progression is known and the response of the control system to a deviation can be evaluated.

The HAZOP follows the conventional method, but common differences encountered with computer-controlled systems include:

**Guidewords**: The usual set of guidewords should be used, but the sequence-related guidewords, such as *sooner* and *later*, and *before* and *after*, become important. Other possibilities are *more* or *less often* and *interrupt*.

**Parameters and deviations**: While considering only meaningful deviations, a sensible selection must be combined with lateral thinking and imaginative suggestions. A computer-controlled system could introduce new parameters such as *data flow*, *data rate*, and *response time*.

**Causes**: As well as the usual HAZOP causes, there may now be additional ones that originate in the computer system, such as the reliability of power supplies, the possibility of complete or partial hardware failure, plus the links and handover sequence. Human factors during interactions with the computer should also be considered.

**Consequences and safeguards**: The evaluation must ask what information is provided to the computer, how it will be interpreted, and how the control system will react. Again, the interaction between the operators and the computer system may enter the analysis.

It is helpful to ask **four key questions** when assessing a developing event:

Does the computer know?
What does the computer do?
Does it tell the operator?
What can/does the operator do?

Clearly the HAZOP study team must have sufficiently detailed knowledge of the intended control and emergency sequences if an accurate evaluation of the consequences is to be made. The team should include both the person responsible for creating the control program and someone who understands the coding process to ensure that there is no uncertainty or ambiguity between the different disciplines involved.

**Actions**: Many actions will concern the draft control sequence, perhaps to provide better control, more checks, further information for the operators, or more alarms. These can benefit from the capacity of a computer to carry out complex checks and actions with speed and reliability. There may also be some actions to correct problems of timing identified in the draft control sequence.

**Reporting**: The normal requirements for good reporting apply. In particular, it is essential that all actions are written so those carrying out the programming or coding will be clear what is required. The report must be clear and meaningful to individuals who were not part of the team, bearing in mind that the implementation will rely not only on those versed in the usual engineering disciplines but will now involve software specialists.

## 10.1.4 Computer HAZOP (CHAZOP) Study

This staged approach is recommended for complex control systems, when there are major hazards or when safety-related functions come under computer control. An enhanced HAZOP is usually carried out first but, if all the necessary information is available, the two studies may overlap. The timing of the second study can be a problem, in that the detailed coding may not be available until late in the overall development.

### 10.1.4.1 Detailed Study of Computer Hardware

Details of the second stage computer HAZOP are given in a Health and Safety Executive (HSE) Research Report,[25] and only a brief summary is given here. If done in full, it is a very comprehensive review of the computer hardware and software. It is very time-consuming, especially for batch processes, and therefore should only be used for high hazard activities or new or unusual processes.

All the normal plant documentation used for the first HAZOP study is needed as well as the detailed design specifications for control schemes and protective systems. For this study, the detailed design specifications for all the computer hardware will also be needed, including cabinet details, I/O configurations, alarm and trip schedules, communication links, watchdogs, backups, power supplies, security considerations, and provisions for software modifications. Many of these will have been identified by HS 2.

The team work through the computer system to build up a picture of how it is intended to work and what will happen if any element fails, identifying all reasonable causes of failure both internal and external to the computer system. In many ways this parallels failure modes and effects analysis (FMEA), by considering all of the failure modes of the key elements in turn. The four key questions used in enhanced HAZOP (see above) will help the team to identify the adequacy of the required response.

This stage of the examination need not be lengthy if equivalent hardware has been considered before for other projects or if there is a consistent, standard policy over hardware, and there are common, tested subroutines in the computer code for actions that recur regularly.

### 10.1.4.2 Detailed HAZOP of Computer Sequences
This follows the conventional HAZOP method of using guidewords, but in this case it applies them to the steps of the control sequence and the computer actions. It is therefore most usually called a CHAZOP. It can obviously only be carried out when the sequence (but probably not the coding) has been fully developed. It can also be applied to the functional steps of a continuously acting control loop.

The sections or nodes for the HAZOP will follow the steps of the control sequence, and as usual a design intention will be built up for each node. But each design intention is now written from the point of view of the computer functions, for example, "Open Valve XSV0001" or "Sense level in Tank T0002."

The conventional HAZOP guidewords are then used to generate deviations such as *no/more/less* signal or *no/more/less/reverse* driven, *high/low/bad* signal, input *other than* expected, *sooner/later* valve movement, etc. As always, the team needs to use all the guidewords creatively to create meaningful deviations.

The guideword *other* can be used to include global effects on the computer system such power failures or environmental impacts, if these have not already been picked up under the earlier guidewords. It can also include key overrides of functions and MOC in control logic, and the important security issues associated with both of these.

For each meaningful deviation, the team then looks for realistic causes and examines the possible consequences, following the conventional HAZOP structure. For significant consequences, the safeguards are identified, and again the four key questions will help the team confirm if these are adequate. The result will be a sequence which has been reviewed systematically for possible deviations at each stage, and therefore will be better able to deal with problems when in operation, neither being unable to recognize what is happening, nor getting stuck in some loop in its internal logic.

For a batch system, this depth of study is very time-consuming but it reflects the fact that the system has to cope with many different circumstances, each providing opportunities for failure or having different consequences that may not have been identified elsewhere. But, despite the time involved, this method has been demonstrated to provide a thorough way of identifying potential problems in the logic and the components of the computer system.

## 10.2 HUMAN FACTORS

The purpose of a HAZOP study is to examine possible deviations from the design intention, to find any previously unconsidered causes of deviations, evaluate their potential consequences, and to then review the relevant safeguards before suggesting appropriate actions. Each of these steps may involve people. This may occur through an error that contributes to a hazardous event or reduces the reliability of a control measure that is intended either to prevent the hazard or limit the consequences. Thus, it is essential that team members take account of human behavior and have a realistic understanding of typical human performance in both normal and abnormal conditions. There are several useful documents[26–28] relating humans and risk. They give many examples of both large and small incidents where human factors played a significant role, describe the main types of human error and how human behavior relates to these, and include guidance on ways to minimize such errors. Indeed, one approach to the management of human failures is described as a "human-HAZOP." There is no doubt that the regulator attaches importance to the qualitative assessment of human failure, backed up where necessary by quantitative assessment. Also there is emphasis on the need to evaluate low-frequency high-consequence events adequately since experience has shown that major process safety incidents are often triggered by human error and organizational failures. This section is intended to provide basic guidance on the human aspects that should be considered in a HAZOP study.

Human behavior falls into three broad patterns. For much of the time, humans operate in a *skill-based mode*, carrying out familiar tasks and actions without having to think consciously about them. This will apply in many industrial operations once they become familiar. For non-standard or less familiar tasks, humans move to a *rule-based mode*—using the available information and trying out a response that seems to fit or has worked in the past. Finally, if nothing easier has worked, they move to a *knowledge-based mode*, seeking further information and trying to find an explanation that will allow a suitable response. Each of these modes is associated with particular types of error. A HAZOP study team needs to be alert to the possibilities of these errors causing or contributing to unwanted outcomes. The team must try to anticipate possible slips, lapses, mistakes, and deliberate violations.

It should be recognized that errors do not occur because people are stupid or incompetent but as a result of fundamental limitations of human performance that are influenced by equipment design features and operational conditions. An HSE guide[26] identifies three contributing aspects— the individual, the job, and the organization. The individual's competence involves skills, personality, attitudes, and risk perception. The job covers the task, workload, environment, display and controls, and procedures. Finally, the organization can affect outcomes through culture, leadership, resources, and communications. However, even accounting for these, it should be recognized that the possibility of human error cannot be absolutely eliminated by training and procedures—these are not adequate control measures for human fallibility.

When carrying out a routine procedure, an operator will mostly work in the skill-based mode. The likely errors here are *slips* or *lapses*. It may be that there is an array of buttons, labeled A–E, to operate similar valves on different vessels. Pressing C when it should be B would be a slip. How easily this might occur will depend on many factors such as the layout and design of the control panel (e.g., where equipment elements from different suppliers have different operating controls or philosophies) as well as external influences such as time pressure and fatigue. The consequence could vary from a trivial loss of material to causing a catastrophic runaway reaction. Clearly this is a cause that the HAZOP team should consider. A lapse might happen in a multistep start-up procedure where, say, after completing step 13 the operator is distracted by a phone call or has to briefly attend to another task. Returning to the sequence it is resumed at step 15 and thereby step 14 is omitted. This may have a trivial consequence; it may be recoverable; but it should be considered by the HAZOP team whenever the consequences matter.

The next level of operation in the hierarchy is rule based. When an uncommon event occurs, humans take the available information to see if it fits some previously experienced or learned rule. The sequence followed is "if the symptoms are X then the problem is Y; if the problem is Y then do Z." More than one rule may be tried if the first does not work. If no rule works, then the knowledge-based mode must be tried. New data must be sought and an attempt made to model the process and use this to select the best actions, improvising in an unfamiliar and possibly critical situation. Not surprisingly mistakes are more likely in these modes. If the knowledge-based mode is called for in a complex system, especially in a critical situation where individuals are highly stressed, the likelihood of successful control and recovery is very low. The HAZOP team must recognize that people under pressure are susceptible to predictable errors due to natural biases within the human cognitive system. People are very bad at recognizing new situations and will tend to jump to hypotheses based on more familiar situations. This can mean that operators will be slow to react to a potentially hazardous mode of operation and assume that the system will, as it usually does, operate safely. People will even try to rationalize weak signals of failure to explain away potential problems; they may even focus inappropriately on evidence that appears to support their assumptions rather than acknowledge that they may be witnessing a new problem.

An example of these modes of behavior—skill based, rule based, and knowledge based—would be a control room operator realizing that during a routine transfer between vessels that the connecting line instrumentation shows a rising pressure—a skill-based level of operation. Applying experience the first rule might be that a valve is closed in the transfer line and this would be immediately checked. If the valve is found to be open, then there is no other obvious cause. With the pressure still rising, further information and a new model are needed. This material has a high melting point—if the operator knows this, then a line blockage may be suspected and then appropriate actions can be tried. Again, knowledge and experience are crucial to raising the chance of a successful intervention.

Finally, the HAZOP team should be alert to possible violations (i.e., deliberate breaches of rules and procedures). There are many possible reasons why violations may occur. It could be to save time, to make the work physically easier, because it simplifies a procedure or seems more efficient. If done during normal, everyday operations, deliberate violations may play a part in eroding safety margins. Where shortcuts in maintenance and calibration tasks, for example, are condoned as accepted practice, the reliability of designed safety measures can be reduced and may one day lead to a major incident. These violations are most likely to take place if employees have the perception that management want corners cut to save time or to achieve the production schedule. Good design of plant and procedures, involvement and education of the operators as well as good management and supervision reduce the likelihood of routine violations, although in an emergency it is possible that irregular steps will be tried.

Within HAZOP study, it is often necessary to assess the likelihood of event frequency. This is usually done by experienced judgment, occasionally by semiquantitative assessment and, rarely, by referral for QRA. These approaches can also be applied to human error. For relatively frequent events, an experience-based approach will work. Estimates may also be derived by task analysis methods using a quantitative Human Reliability methodology but this takes considerable effort and requires considerable expertise. At the intermediate level of estimation, there are some helpful observations from within the nuclear industry,[29] and it is useful if the team leader or at least one member of the team has knowledge of these documents. They suggest that there is no task, however simple, for which the failure rate is zero. For the simplest task listed, the selection of a key switch operation rather than a non-key one, the quoted error rate is 1 in $10^4$—this implies that no task is error free. So a study team should never assume that a problem can be eliminated completely by an action that relies entirely upon an operator. At the other extreme, for example, the high-stress situation of large loss of coolant in a nuclear reactor, the probability for "operator fails to act correctly in the first 60 s" is 1. That is it should be assumed that there is no chance at all of correct remedial actions in that time. The situation does not improve greatly over the next 5 min and is not negligible several hours later. Another source of human error can occur at a shift handover where communication and records of previous actions may be poor and an error rate of 1 in 10 is quoted for "personnel on different work shift failing to check the condition of the hardware unless required by checklist."

While these are useful guidelines, it is important to recognize the many other factors that influence human error rates. A comprehensive set of performance-influencing factors (PIFs) has been established.[30] These include training, control panel design, competence and motivation, environment, level of demand and suddenness of onset of events, management attitude to safety, procedures, and communications. There are many more. Understanding these may influence a HAZOP team's suggestions for action. In a modern computer-controlled plant, it can be easy to add an alarm but if this is to be done it must be within the overall design of the alarm and trip system so that the operator is not subjected to alarm and/or mental overload when a major event occurs. When an individual is overloaded with information, they are less likely to separate the critical, top-level information from the unimportant and the trivial, resulting in either inaction or the wrong action. Another state is mind set where the individual uses the information to create an initial, but erroneous, scenario and rejects critical information which shows it to be incorrect.

The HSE document, Identifying Human Failures,[31] gives a list (in the following table) of failure types in the form of HAZOP style guidewords which may be used in the search for human error leading to a deviation.

| Action errors | Checking errors |
|---|---|
| A1: Operation too long/short | C1: Check omitted |
| A2: Operation mistimed | C2: Check incomplete |
| A3: Operation in wrong direction | C3: Right check on wrong object |
| A4: Operation too little/too much | C4: Wrong check on right object |
| A5: Operation too fast/too slow | C5: Check too early/too late |
| A6: Misalign | |
| A7: Right operation on wrong object | *Information retrieval errors* |
| A8: Wrong operation on right object | R1: Information not obtained |
| A9: Operation omitted | R2: Wrong information obtained |
| A10: Operation incomplete | R3: Information retrieval incomplete |
| A11: Operation too early/too late | R4: Information incorrectly interpreted |
| *Selection errors* | *Information communication errors* |
| S1: Selection omitted | I1: Information not communicated |
| S2: Wrong selection made | I2: Wrong information communicated |
| | I3: Information communication incomplete |
| *Violations* | I4: Information communication unclear |
| V1: Deliberate actions | |

A HAZOP study team would seldom find it necessary to systematically examine all of these possible deviations. In many operations and procedures, the use of appropriate guidewords, which may be problem specific, will help the team decide which of the possible deviations could lead to potential problems. It is unlikely that all of these deviations would be found using just the conventional combinations of guidewords and parameters.

The example of a HAZOP study of a procedure (Appendix 5) shows some of the ways that human factors may be identified by the team.

In summary, all HAZOP study teams need to be aware of the potential for human error to generate causes and to influence consequences. They need to use the present understanding of human behavior, influencing factors, and the typical probabilities for different types of error. It is also good practice to examine the design of control screens from the perspective of the operator. This will reveal design inadequacies such as when separate elements that should be monitored as part of a routine task are actually presented on separate screens. Such arrangements add workload and complexity and introduce opportunities for confusion and error. In formulating actions, they should consider the required level of human behavior—the *skill-based*, the *rule-based*, or the *knowledge-based mode*—and the actions should reflect the needs for further diagnostics, training, second line of supervision, or simply an addition in a standard operating procedure (SOP) as illustrated in Appendix 5.

## 10.3 LINKING HAZOP STUDIES TO LOPA

Layer of protection analysis—LOPA—is a widely used technique to determine the level of protection needed to provide adequate safeguards against major hazards that could arise on a plant or process. The method, developed in the early 1990s, is well documented[32] and is accepted by regulators in many countries as an appropriate method of analyzing identified hazards and assessing if sufficient protective systems are in place to achieve a tolerable risk.[33] The first step in LOPA is the classification of the severity of hazard consequences if this has not already been carried out by a Hazard Identification process such as HAZOP. For the major consequences such as injury, fatality, or major accident to the environment, the magnitude of the ultimate, unmitigated consequences (a scenario) is estimated, then a maximum tolerable frequency is assigned from either the company standards or from publically available suggested levels.[33] The frequency of the initiating cause(s) is estimated using either a generic value for this type of event such as control system failure or human error or, preferably, by fault tree analysis of the expected sequence for the individual event. Other factors such as "time at risk" for hazards which exist for part of the time and "conditional modifiers" such as probability of ignition are considered. It is then possible, taking into account all *independent* protection layers, to estimate the frequency of occurrence of the scenario, that is, the frequency of the top event. Comparison with the target tolerable frequency shows whether the protection is adequate and, if not, the magnitude of the necessary improvements. Where instrumented safety systems are used, their reliability (PFD) is often expressed as an SIL. The further measures to achieve a tolerable frequency may include addition of an SIS—a system designed and evaluated from sensor through control loop to actuators—to have a demonstrable PFD at levels such as SIL1 (a PFD between 1% and 10%) or SIL2 (a PFD between 0.1% and 1%). LOPA may be used as a part of HS 2 to ensure that the PFD of each SIS can be covered in the detailed design. Alternatively, or additionally, it can follow HS 3, especially where this involves a full HAZOP study.

As the prime aim of HAZOP study is the identification of hazardous events and evaluation of the consequences, it can clearly link to a LOPA study, since it produces deviations, causes, consequences, and safeguards which feed directly into LOPA. To realize the potential of this link, it is important that the team fully evaluate and record in detail the consequences for each cause of each deviation as well as the safeguards already present in the design. Thus, any possibility of severe injuries or fatalities, of major fire, explosion and toxic releases, or of substantial plant damage and disruption, need to be recorded in the consequence column of the HAZOP report. This highlights scenarios that should be included in the LOPA study. An additional useful item is the teams' view of the likely frequency of the event.

Two outcomes are common during the HAZOP study. If the scenario has already been considered at the HS 2 stage, then the team can immediately review the design and recommended safeguards to confirm their adequacy. If the scenario is new then, as a potential major event, it will need to be examined using the LOPA method-ology. This requires a LOPA study to follow the HAZOP study either as a resumption of the earlier one or a new study. To be effective the HAZOP team must be on the lookout for potential major con-sequences, even when these are anticipated to be very-low-frequency events. The ultimate consequence should be clearly recorded so that when the records are reviewed they stand out as needing review in the LOPA study. All possible causes should be recorded as these are essential inputs to the LOPA study. Similarly, the existing safeguards should be clearly recorded as they may serve as independent layers of protection against a developing event.

Thus, LOPA and HAZOP study are natural partners in identifying hazards, determining whether existing safeguards are adequate and, if not, specifying the additional levels of protection that are needed for each possible cause of the event. However, if the full benefits are to be realized, it is essential that the HAZOP study has been planned to provide optimal information for the LOPA study.

# CHAPTER 11

# Specific Applications of HAZOP

The main uses of HAZOP study in the process industries are for new designs, processes, and operations, both continuous and batch, and for modification and reuse of existing plant and processes. This chapter comments on the special aspects of these uses and also considers some other, less frequent, applications.

## 11.1 MODIFICATION OF EXISTING OPERATIONS

It is important to have in place a procedure for the management of change (MOC) to ensure that all modifications are reviewed before any variation in plant, process, or operation is made. The review should recommend an appropriate method of hazard identification. Where there are significant hazards, this may be a HAZOP study. The company MOC procedure should include criteria for deciding if a HAZOP study should be done.

A very wide view should be taken as to what is a "modification." Anything which changes a plant or a process in any way must be treated as a modification. Such changes could be to materials, catalysts, solvents, conditions, sequences, quantities, procedures, software, and so on.

When a HAZOP study is used for a modification, the basic principles of HAZOP are retained and applied. For major modifications, the study follows the steps taken for a new design. For small modifications, it is possible to proceed more quickly, using a smaller team and combining some of the roles within the team—for example, a member may act as the scribe and it may even be acceptable for the leader to have another role. It is particularly important to have the operating team represented within the HAZOP study team.

If the system was previously studied by HAZOP, then the original report may provide a useful starting point. There can be problems with defining the boundaries for the study since it is unlikely that the whole operation will be reviewed. The boundaries may have to be some distance from the point of the modification to ensure that all

HAZOP: Guide to Best Practice. DOI: http://dx.doi.org/10.1016/B978-0-323-39460-4.00011-6

relevant causes and consequences are considered. The boundaries should be agreed between the project team and the HAZOP leader with the leader given the authority to extend the boundaries, if felt necessary, during the study.

## 11.2 REPEAT DESIGNS—HAZOP-BY-DIFFERENCE

In some branches of the process industry, it is commonplace to install designs which are essentially the same as an earlier installation or which are made up of standard units, varying only in size from other installations. In these cases, it may be possible to do an effective hazard identification study by detailed comparison to an earlier, full HAZOP study, concentrating on any differences from the previous case. When this method is used, the team must be particularly aware of any variations in size, site, services, and interfaces with other plant.

## 11.3 PERIODIC HAZARD STUDIES AND THE HAZOP OF AN EXISTING PLANT

Periodic hazard studies are process hazard analyses to ensure a process plant continues to operate and be monitored to appropriate SHE standards throughout its life. While MOC reviews provide a record of incremental changes over a period of time, it may become necessary to review a system as a whole, particularly when the multiple changes may interact adversely with each other.

Such a review is particularly important if any changes to operating procedures, feeds, or products and/or modifications have been made. The requirement for such periodic studies can be legal—for example, OSHA—or company policy as best practice. There are several techniques available for such studies, including the retrospective use of Hazard Studies 1 and 2. HAZOP study should be considered as a preferred approach if the following have occurred:

- major incidents;
- many modifications;
- the original studies were inadequate;
- significant design deficiencies have been revealed;
- the plant has not run smoothly.

HAZOP is necessarily more time-consuming than most of the alternative techniques, but has the advantage of a comprehensive outcome.

Other techniques[2] can be valuable in identifying key issues and the need (or otherwise) for more detailed studies such as HAZOP. The choice of method will depend among other factors on the available experience base, the sophistication of the process, and regulatory requirements.

Whichever technique is used, it is important that target dates are set for completion of actions and for review of progress and subsequent periodic studies. Proper action progressing and specific periods for subsequent study may be a legal requirement where such studies are mandated.

## 11.4 OPERATING PROCEDURES

The HAZOP methodology for an operating procedure is essentially the same as for a batch process. Such a detailed study is normally only applied to critical procedures. It requires a well-defined procedure to be available, including all significant steps and actions, an up-to-date P&ID and, ideally, structural drawings to locate valve positions. For existing plant, a tour of the process area is recommended. In preplanning, it has to be decided whether the HAZOP of the procedure comes before or after the HAZOP of the process—the normal case would be for the process HAZOP study to be done first. Also, the study must not degenerate into a procedure-writing meeting. The team composition must be correctly balanced to get the best results and must include some members who are familiar with the process, including an operator. Before the start of the study, the procedure is reviewed for clarity and the aims of the study defined. Normally, these are to identify potential hazards, operability problems, and environmental problems which may result from deviations from the procedure, especially those due to human factors.

The actual analysis follows the batch HAZOP methodology, working through the procedure stage by stage. Each stage, which may consist of a number of individual actions, is examined using the guidewords to prompt the team members to suggest meaningful deviations which are then analyzed in the usual way. In addition to the standard guidewords, "out of sequence" and "missing" can be productive. "Missing" is interpreted to mean that a step is missing from the procedure at or just before the stage examined—although such deviations could equally well be found using the guideword "no." In the list of parameters, the phrase "complete the step" can be used to good effect, as it combines meaningfully with the guidewords "no," "more," "less," "reverse," "part of," "as well as," "out of sequence," and "missing." An unusual but occasionally useful question is "verification of success"— how is it known that a valve is closed or that a vessel is depressurized. This last point is very important in pigging and filter operations.

A major difference from process studies is that many of the causes of deviations are related to human factors. These may be of omission or commission. The importance of Human Factors in HAZOP studies has been considered in Section 10.2. Other possible causes include poorly written procedures, difficulties caused by poor layout, bad lighting, parameter indicators with limited or poor ranges, or too many alarms. The latter, a cause of information overload, is a topic of concern and advice is available.[9] In assessing safeguards, a reasonable allowance can be made for the presence of the operator if close involvement with the system allows for the possibility of immediate detection and correction of the deviation. Experience of the actual conditions and the style of operation is important when making such a judgment about human factors.

Actions may suggest a change in the procedure but need not be limited to this option; instrument or equipment modifications should be recommended if they offer the best solution to a problem.

## 11.5 PILOT PLANT AND LABORATORY OPERATIONS

Pilot plants and laboratories typically differ from full-scale plants and processes by their smaller scale, diversity and greater degree of human interaction. Nevertheless, there are many exceptions. Some refinery or petrochemical pilot plants can dwarf full-scale fine chemical plants. Pilot plants built to test scale-up may be single continuous stream plants and highly automated. However, the hazard study approach for all pilot plants, semi-technical plants, and experimental plants follows the same pattern. A preliminary hazard analysis—for example, Hazard Studies 1 and 2 (see Chapter 2)—should be carried out, and if the potential for significant process hazards is identified then a HAZOP study can be recommended. The same HAZOP methodology for continuous processes or batch processes, as described earlier, is used but with greater emphasis placed on the process and design uncertainty and human factors.

The design intent and limitations of the pilot plant or laboratory should be clearly defined. Constraints need to be established on the experiments, materials, and process conditions allowed in the pilot plant or laboratory. A system needs to be established which registers and evaluates new experiments. If an experiment falls outside the established constraints, then the experiment should be subjected to a preliminary hazard analysis which decides whether a HAZOP study is required. The preliminary hazard analysis should also establish whether extra studies are required—for example, reaction stability,[34] reactor relief requirements, emergency measures, and occupational health. Changes to the building, laboratory, plant, process, or equipment are covered by a modification procedure, as discussed in Section 11.1, page 77.

It is recommended that the HAZOP study of large continuous pilot plants initially studies the process as a continuous one (see Section 5.3.1, page 37). During the HAZOP study, extra emphasis is required on the possible inadequacies of the design. For example, a heater control may not just fail but the heater may be grossly over- or under-sized as a result of uncertainties in the process stream properties. Since starting, stopping, and aborting experimental runs is a regular feature, the process steps involved in these operations should also be subjected to a batch HAZOP study (see Section 5.3.2, page 39).

For multifunctional experimental plants (typically batch or semi-batch processes), it is recommended that a typical process is selected and subjected to a HAZOP study. The other processes are then hazard-studied in as far as they deviate from the established process in conditions (e.g., temperature, pressure, and concentration), materials (e.g., additives and solvents), and process steps. For all types of unit, the operating bands must be clearly established at the start.

For typical research laboratories, support systems should be subjected to a HAZOP study if they are critical for the security, safety, health, or environment of the laboratory. For example, the ventilation and extraction systems for a laboratory designed to handle highly toxic materials or biologically active agents should be effective and reliable— a HAZOP study of these systems should identify the potential hazards. A HAZOP study of gas supplies identifies hazards caused by failure of the equipment or operation not covered by standard designs.

For pilot plant and laboratory HAZOP studies, an appropriate team is assembled. A good team might involve the laboratory manager, operating technician or laboratory analyst, equipment specialist, maintenance manager or technician, and HAZOP leader. If the purpose is scale-up, then the respective idea developer—for example, chemist, physicist, and engineer—and a technical representative of the subsequent development stage are useful additions.

Due to the greater degree of human interaction, a greater emphasis is placed on the skill, knowledge, training, and experience of the operating technician or laboratory analyst. Involving these in the HAZOP study ensures relevance, simplifies training, and aids motivation, as well as improving the documentation of procedures.

A HAZOP study of proprietary equipment—for example, analytical test equipment—is generally not required provided the equipment has been subjected to a risk analysis by the manufacture and is installed and used only as intended by the manufacturer.

While no laboratory experiment is too small to be hazard-studied, the benefits of HAZOP study are more likely to be achieved if there is a potential for fire, explosion, significant release of a hazardous material, or other major loss. A HAZOP carried out at this stage can help to ensure that hazards are addressed at an early stage in the development of the process.

## 11.6 DRAINS, VENTS, AND OTHER INTERCONNECTIONS BETWEEN PLANTS

The vent, relief, and drains systems often link many pieces of equipment, sometimes different plants, through a common piping network. The individual pieces of equipment may operate at significantly different pressures, some equipment may be starting up while others may be shut down, and some of the fluids may be mutually incompatible. The design of these systems is often complex to reduce release to the environment and may be spread over a number of process P&IDs or split between process P&IDs and a separate set of vent and drain P&IDs.

The HAZOP therefore requires special skills:

- interface management between P&IDs (of which there may be many);
- analysis of fluid incompatibilities;
- analysis of the potential for simultaneous releases (particularly vent and relief systems);
- assessment of the potential for dynamic, static, or other induced chokes.

The interface management requires at a minimum that all interfaces (lines entering and leaving a P&ID) are positively identified on all drawings and that there are no mismatches or exclusions. If there is a set of vent and drain P&IDs, each interface on the vent or drain P&ID should be labeled with a fluid description. The following data may be recorded:

| | Condition in Equipment | Condition in Vent/Drain |
|---|---|---|
| Flow | ✓ | ✓ |
| Phase | ✓ | ✓ |
| Pressure | ✓ | – |
| Temperature | ✓ | ✓ |
| Potential incompatibilities (solids, water, acid/alkali) | ✓ | ✓ |

If the vent and drain are included on the process P&IDs and there are many drawings, consideration should be given to preparing a special interface drawing to link all process P&IDs onto one sheet. If this is not possible, the interfaces can be labeled as above and the vent or drain treated as a system. The main parameters and guidewords are:

| | |
|---|---|
| Flow | More/reverse/no |
| Pressure | More |
| Temperature | Higher/lower |
| Phase | Change |

However, there are likely to be secondary issues such as:

| | |
|---|---|
| Pressure drop | High |
| Line drainage | No/less |
| Dynamic choke | More |
| Static chokes due to debris | More |
| Imposed back pressure on the relief valve | More |
| Reaction forces | High |
| Isolation standards | Less |
| Material compatibility | Less |

A study such as the relief and blowdown review examines the dynamics within a subsection and the total system, but this does not eliminate the need to examine the total system with a HAZOP study.

## 11.7 COMMISSIONING AND DECOMMISSIONING

Commissioning and decommissioning occurs only once on any process and the issues are often unique. The problems of commissioning are usually dealt with during the main HAZOP study, either by inclusion as a parameter under the guideword OTHER in simple cases or by full examination in more complex cases. They usually have to be treated as a sequential operation rather than as part of a continuous process. Decommissioning is seldom considered during early project stages; it can be a complex process which merits its own HAZOP study before it is undertaken.

The main features of commissioning are:

• removal of construction debris;
• purging;
• test runs for equipment.

Where the process is very critical or involves complex or high-cost machinery such as major compressors, the HAZOP procedure can be used to follow the cleaning process, for example, to:

• verify that no debris is moved from dirty to clean systems;
• identify where debris may lodge and/or block restrictions (valves, flow meters, instruments);
• ensure that cleaning proceeds from small to larger piping and not the reverse.

The purging routes can be managed in a similar manner but, obviously, the rule is to ensure purging proceeds in one direction only.

Test running equipment has the potential to operate outside the normal envelope. Fluid velocities may be higher (or lower) than normal, the test fluid may have a different density, viscosity, or temperature and, in the case of gases, a different ratio of specific heats. If water is used instead of a lower density fluid, the static and dynamic heads and power draws may be excessive and the static loads on piping may be higher. If air is used for test-running compressors, there may be seal problems, horsepower limits, and high discharge temperatures.

When HAZOP is used in connection with commissioning, it is necessary to select suitable parameters in order to develop meaningful deviations. Some examples are given below:

| | |
|---|---|
| Density | Higher/lower |
| Molecular weight | Higher/lower |
| Pressure ratio | Higher/lower |
| Power demand | Higher/lower |
| Gamma | Higher |
| Noise | Higher |
| Debris | Some/more of |
| Contamination | Oxygen/inerts—source of/disposal of |
| Process contaminants | As well as |
| Water | Consequences of/formation of |
| Cleanliness | More/less |
| Pressure | Over/under |
| Load/stress | Higher |
| Other | Projectiles: more/less/velocity |
| Other | Ice/mass balance/static load |

In the case of decommissioning—leading to demolition—the sequence with which the equipment is decontaminated is followed by the HAZOP procedure. Furthermore, the procedure can follow potential issues associated with:

* catalysts plus contamination reactivity/deactivation;
* pockets or potential traps;
* special procedures for the demolition contractor.

Ultimately, as much of the equipment as possible should be recycled and the manner in which it is decommissioned may affect the demolition. There may be some form of process waste (even residual working inventories) which has to be processed further elsewhere. This requires an abnormal operation which should be studied in detail.

## 11.8 START-UP AND SHUTDOWN

In general a process plant is designed for steady-state operation. The piping configurations and instruments are designed for that objective. But there are also two dynamic modes which also must be considered—start-up and shutdown—especially important since many recorded incidents have occurred during these phases.

The main purpose of a HAZOP Study is the analysis of possible deviations outwith the design envelope during *steady-state operation* of a plant or process and has not handled Start-Up and Shutdown adequately. The structure of a Start-Up and Shutdown will be in the form of a procedure or set of steps which start at the introduction of inventory and end with the specification product from that step. The Start-Up and Shutdown HAZOPs are therefore best studied as a Procedural HAZOP (Appendix 5).

During these operations, pressures and temperature, and hence compositions of the process fluids, may be well outside the steady-state range or even outside the specification of the materials of construction. Also, as the illustration in Appendix 5 shows, if a step is missed or is done out of sequence there is the potential for a deviation.

This section must be treated as an overview as each unit operation must be treated separately. It should be noted that control is designed for steady-state operation and not necessarily for the dynamics of Start-Up and Shutdown. Attention must be paid to the control of a potentially unsteady state as well as specific flows only experienced during these operations.

### 11.8.1 Start-Up

Start-up is a stepwise process. The division or specification of the nodes is case specific and best defined by the Facilitator and team; they may well be different from that of a steady-state study. Appendix 5 gives some ideas.

Start-up begins with a pressurizing cycle for that step followed by a conditioning phase ending in specification product leaving that step.

During a pressurizing cycle, the deviations:

- pressure low;
- temperature low;
- velocity high

can be used to cover potential problems related to vapor liquid equilibrium data. There can often be problems due to effects in vent and relief systems. Low temperature may create fluid handling issues due to viscosity, particularly in pockets or traps, or metallurgical issues due to the properties of the materials of construction. Also during start-up, inerting materials such as nitrogen (as well as) may create difficulties with condensation.

Consideration should be given to the various steps in the conditioning operations using guidewords such as:

- out of sequence;
- incomplete;
- too early or too late;
- too fast or too slow;
- step missed out;
- nonsteady-state control.

These may be treated as generic and applicable to all unit operations. As the example in Appendix 5 shows, these are appropriate to any procedural HAZOP study. As the Start-Up is followed until steady-state operation is achieved, attention must be paid to the product composition at any part of the process using guidewords:

- as well as;
- other than.

Some unit operations may require a second set of guidewords. For a crystallization operation, the following might be appropriate:

- too small or too large (crystal size);
- mass balance.

Attention must be given to how the off-specification materials produced during the Start-Up can be stored, reduced, reused, or recycled, bearing in mind the potential upset that might occur if off-specification materials enter the downstream section of the process. Guidewords such as the following can be used:

- How much?
- Where?
- As well as
- Metallurgy.

### 11.8.1.1 Special Start-Up Conditions

It is important that the team follows the dynamics of the process during Start-Up. There are many potential hazards that must be considered and for which no listing would be adequate. As an example the start-up of a refrigeration compressor which also acts as a heat pump will require a special recycle loop which could be operated manually for only an hour. The loop is not required for steady state but if not fitted the compressor cannot be started up if "flash gas" is normally used for the heat pump cycle. Likewise during start-up, it may be necessary to circumnavigate the flammability diagram bearing in mind that it is both pressure and temperature sensitive.

It should be noted that any Start-Up may require a total or partial Shutdown. There will be a number of upsets which will require the identification of a "Safe Holding Position" which may not necessarily be the total shutdown and could be at a stable part of the Start-Up cycle. The team should consciously address this Safe Holding Position in the procedure from where the restart could be attempted.

The team should select an appropriate set of Guidewords.

### 11.8.2 Shutdown

Shutdown is less of a stepwise process and is not necessarily the reverse of start-up. The division or specification of the nodes is case specific and best defined by the Facilitator and team; they may well be different from that of a steady-state study. Appendix 5 gives some ideas.

Inventories must be run down to an operating minimum. Then the process is depressured and residual and often contaminated inventory recovered. Once again guidewords such as follows may be appropriate:

- low temperature;
- low pressure;
- increased viscosity;
- phase change (phase change could be ice);
- vacuum.

Consideration must be given to how and where the residual contaminated inventory might be stored, recovered for reuse, and recycled during a turn-around or destroyed when the process is shut down permanently and decommissioned (see Section 11.9).

Guidewords such as the following can be used:

- Where?
- How much?

- Too much (the volumes of contaminated inventory at the final shutdown may be critical)
- Other than
- As well as
- Metallurgy

### 11.8.2.1 Special Shutdown Conditions
As with start-up it might be necessary to operate a flash gas recycle line or to circumnavigate a flammability envelope.

### 11.8.2.2 Emergency Shutdown
Emergency shutdown by a SIS can be tracked through the Cause and Effects Diagrams to ensure that the sequence is correct. Even if correct hazards could still occur due to faulty operation of the SIS. Guidewords such as the following should be applied to each operation:

- incomplete operation;
- partial operation;
- nonoperation;
- reverse operation.

## 11.9 CONSTRUCTION AND DEMOLITION

Construction is not a steady-state condition, and equipment may be delivered out of sequence; furthermore, there may be the need to impose unusual loads—for example, in the hydraulic testing of a steam main or gas main. The issues are likely to be the following:

- sequence of equipment (out of sequence);
- access for lifting equipment into place;
- loads on piping/foundation.

Where construction is undertaken close to an operating plant, the potential for interaction may need to be considered.

Demolition is not the reverse of construction and contains its own SHE risks. Safety issues involve access and overhead/underground operations. Health hazards include toxic/flammable contamination, possible fires, and asphyxiation; some of these hazards are produced in the demolition process—for example, in the hot cutting of materials. Environmental issues could involve the disposal of lagging, spent catalysts, residues found in the equipment, and possible spills on site.

The restoration of the site for other uses is influenced by the mode of demolition if it does not capture all residues. The classification of materials for recycling and the verification of cleanliness are equally important.

Specialist study methods have been developed for construction and demolition activities using checklist and/or HAZOP study approaches.

## 11.10 CONTRACT OPERATIONS

It is important to define within the contract all aspects of any planned HAZOP study, including the extent to which operability problems are sought, the responsibility for cost of change, and the control of team membership. The latter may be a problem if clients request several places in the HAZOP team in addition to the contractors and representatives of licensors. The team leader should still aim to keep the team number below 10 and should consider the responsibilities of each prospective team member to ensure their presence is needed for the sections being discussed.

### 11.10.1 Advice to Users of Contractors

Where a company wishes to use a contractor to carry out a HAZOP study for systems on process plant, reference may be made to this guide in any invitation to bid. It should be made clear who is responsible for the HAZOP study, including closing out all of the actions. Contractors may have their own corporate HAZOP method document (procedure or standard) ready for a client to review and comment on before contracts are signed. If this is either more or less than required, it must be made clear prior to signing. Comments should make it clear whether and where the scope is to be changed. Any changes suggested to the way in which a contractor does the work, after contract award, may be grounds for a contract change. An alternative approach is to give the company HAZOP study procedure to the contractor so that the work can be priced to meet the specified requirements.

In some countries, the contractor may have a statutory obligation with respect to the health and safety requirements of their deliverables and so it may not be practical to reduce the scope of work for the HAZOP study.

When the study is done it may happen that the team is confronted by a package unit which is defined only as a box. The team must make every effort to find the correct P&IDs and to study them in the proper manner. Some of the P&IDs may not follow the same style or give the full details found on process P&IDs.

## 11.10.2 Advice to Contractors

It is good practice to have a corporate procedure or specification which covers the preparing, running, reporting, and following up a HAZOP study and to state how it relates to this guide to best practice. Such a procedure should cover the training requirements and qualifications stipulated for the HAZOP team members, particularly the leader. When contractors bid for a project, it is important that their corporate procedure is submitted, or is referred to, in the contractor's bid. Once it has been given to the relevant client for review (confidentially if necessary), it serves as a benchmark for what will actually be done on the job. Any changes to this procedure required by the client should be documented in a job-specific procedure, approved by contractor and client, for that project. The scope of the systems to be studied should be defined, and any sections not included must be clearly identified.

Where an external consultant is used to lead the team, it is important to make arrangements for access after the HAZOP is completed to resolve any uncertainties which may arise.

# CHAPTER 12

# Factors for a Successful HAZOP Study

There are a number of pitfalls in the HAZOP process which must be addressed and eliminated throughout the study process; before, during, and after. Those listed in this chapter are some of the more common ones that may affect the quality and value of the study. The listing is not comprehensive but serves to indicate the detail required to achieve a good study. Further advice is given by Kletz.[1]

## 12.1 THROUGHOUT THE STUDY

- The HAZOP study should be an integral part of an overall SMS that includes the other appropriate studies described in Chapter 2.
- The process must have the full backing and support of senior management.

## 12.2 BEFORE THE STUDY

- The study must be initiated by a person who has authority and who will also receive and implement the actions. If the person does not have authority and the actions are not implemented, the study is a waste of time.
- The design must be well developed and "firm"—that is, the sections examined are not being simultaneously developed. In the case of a modification, the P&IDs must be verified as "as built" with the changes highlighted within a "cloud." A study cannot be carried out on a partly-developed design as the subsequent changes will undermine the HAZOP study. Freezing of the P&IDs is critical to a study. Also, the drawings must be well prepared. The drawings are the record of what was studied, and if they are inaccurate or incomplete the HAZOP study is worthless. Equally, the study must not be delayed too long as the options for change will become very limited. Premature study where the P&IDs are still not finalized is wasteful of time and effort. Equally, the freeze of the P&IDs stops further

HAZOP: Guide to Best Practice. DOI: http://dx.doi.org/10.1016/B978-0-323-39460-4.00012-8

development of the drawings and may hinder further design work. The balance is a careful judgment.

- A skilled and suitably experienced team leader should be chosen.
- The leader must be given a clear scope, objectives, and terms of reference from the initiator for this study (including delivery date and recipient); if this is not done, the study may be incomplete in some aspects and not fulfill the requirements of the initiator.
- The leader should choose a route plan (see Section 5.3) through the P&IDs to ensure that all necessary sections are covered effectively with special care taken at branches and interfaces with services. The route should be clearly defined with well thought out starts and ends.
- The study should not be required to make project decisions; nor should the design team adopt the approach of "leave it to the HAZOP study to decide what should be done!" If a problem is known, then it needs to be addressed during the design.
- The study team must be balanced and well chosen to combine knowledge and experience. A study group that is drawn entirely from the project team will not be capable of critical creative design review. Equally, a team which has no operations input may lack objectivity.
- The team must be given adequate notice of the study so that they can carry out their own preparation in readiness for the study itself. This may require some preliminary reading of any relevant hazard databases and analysis of the P&IDs.
- The extent to which problems are evaluated, ranked, and solved should be defined.

## 12.3 DURING THE STUDY

There are a number of important factors for success during the actual study process:

- The team must be motivated, committed, and have adequate time to complete the examination.
- Team continuity is important—only essential variations and substitutions should be accepted (see Section 5.2.3).
- The boundary of the study must be clearly analyzed and studied. A change on item "one" may have an effect on item "two." The item may be two different processes or an operation upstream or downstream on the same process. If the potential impact is not perceived correctly, the boundary may be placed wrongly.

- The boundary of a study on a modification is equally complex—a change in the temperature of a reactor may affect the by-product spectrum and have a more far-reaching impact than the immediate modification.
- A clear description, design intention, and design envelope must be given to every section or stage examined.
- The study uses a creative thought process. If it becomes a mechanistic process and simply works through a checklist, or if fatigue sets in, the study must be halted and restarted when the team is refreshed.
- Proposing, developing, and finalizing actions is the responsibility of the team, not the leader.
- Each action must be relevant, clearly defined, and worded with no ambiguity. The person who follows up the action may not have been at the meeting and could waste time and effort if there is a misunderstanding.
- The study must accept a flexible approach to actions. Not all actions are centered on hardware changes—procedural changes may be more effective.
- The study team members must be aware that some problems ranked and identified during the study may be caused by human factors.
- There are potential pitfalls, which must be treated individually, when planning the route around branched systems. These branches may be recycle lines, junctions in the process, or vents and drains.

## 12.4 AFTER THE STUDY

- Every action raised must be analyzed and answered accurately.
- Many of the actions raised will require no further change but all must be signed off as "accepted" for action or no action, as described in Sections 6.4 and 6.5.
- Actions that require a positive change should be subject to an MOC process (which may require a new HAZOP of the change) and put into a tracking register.
- Action reply sheets should be clearly linked to the original study reports, including the reference number, the node, the intent, and the deviation. This greatly helps follow-up and audit. They should also reference any calculations carried out.

# The Guideword-First Approach to HAZOP

In Chapter 4, a detailed description is given of the parameter-first approach to HAZOP study, perhaps the most widely used approach particularly for continuous processes. An alternative is the guideword-first approach, which is also widely used, particularly for batch processes. This was the approach used in many early studies. The two approaches do not differ in any basic essential, only in the order in which deviations are developed and analyzed. The guideword-first sequence is shown in Figure A1.1.

In the parameter-first approach, a parameter is chosen and then considered in combination with all of the guidewords that give a meaningful deviation. Thus, the parameter flow might be combined with the guidewords "none," "more," "less," "reverse," and "elsewhere." Then, another parameter such as "pressure" or "temperature" might be taken, each one being combined with "more" and "less."

In the guideword-first approach, the guideword "none" is taken first and combined with "flow" (plus any other parameters which sensibly combine with it). Next, the guideword "more" is combined with the parameter "flow," then "pressure" and "temperature." These three parameters are taken in turn with the guideword "lower." Table A1.1 contrasts the examination sequence under the two approaches.

With a good leader, the only difference between the two approaches is the order in which deviations are considered. In these circumstances, the choice of the approach is no more than a matter of preference or convention.

One advantage of the parameter-first approach is that all aspects of a parameter are taken together instead of being interspersed between deviations which involve other parameters. Therefore, all possible deviations associated with flow are sought before another parameter is considered. This is well suited to the main parameters of continuous processes.

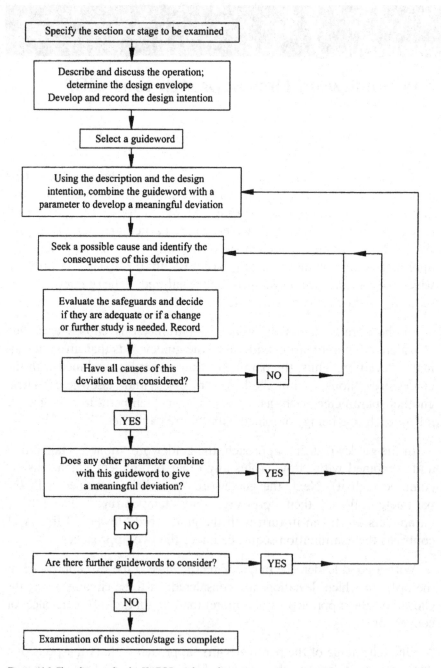

*Figure A1.1 Flow diagram for the HAZOP analysis of a section or stage of an operation—the guideword-first approach.*

| Table A1.1 A comparison of the order of considering deviations with the guideword-first and parameter-first methods | |
|---|---|
| **Guideword-First Method** | **Parameter-First Method** |
| No flow | No flow |
| More flow | More flow |
| More pressure | Lower flow |
| More temperature | Reverse flow |
| Lower flow | Flow elsewhere |
| Lower pressure | More pressure |
| Lower temperature | Lower pressure |
| Reverse flow | More temperature |
| ............... | Lower temperature |
| Flow elsewhere | ............... |
| ............... | ............... |

However, it can be argued that creativity is better encouraged by the guideword-first approach, particularly for the later guidewords such as "part of," "as well as," and "other than." The reason for this is that while appropriate guidewords are easily selected, it is a greater challenge to get a fully comprehensive and imaginative coverage of the parameters. If the parameter-first approach is used, there is a tendency to list the parameters at the start of the analysis and, when the original list is exhausted, to move on to the next section without a final search for any more parameters. This is not good practice.

Consequently, the parameter-first approach may provide convenience, but it demands a greater understanding and application by the team leader and the team members if the best results are to be obtained.

# The Use of Checklists Within HAZOP Study

The use of checklists is contrary to the principles of the HAZOP study method but for an experienced leader, a carefully constructed checklist for a commonly encountered unit can be a helpful way of ensuring that less common deviations are fully considered. With less experience, however, there is a danger that these become a substitute for the creative thinking and analysis from first principles that are essential characteristics of a good HAZOP study.

The examples given here illustrate two ways in which checklists may be used. Table A2.1 lists some aspects which might be considered under the guideword "other (other than)." Table A2.2 gives checklists for vessels and for vents/drains. Neither is comprehensive, nor will all the items in the list need consideration in every analysis. Thus, it is recommended that if lists are used at all they are the responsibility of the team leader who brings out items from the list which might apply in the circumstances of the study. Furthermore, it is important that such lists are customized and expanded to ensure relevance to each industry and application.

**Table A2.1 An illustrative parameter list for the guideword "other (other than)"**

| |
|---|
| Maintenance |
| Testing and calibration |
| Instrumentation, alarms, and trips |
| Hold conditions |
| Services |
| Relief |
| Sampling |
| Static |
| Start-up and shutdown |
| Low and high rate running |
| Nonstandard batches |
| Manual operation of automatic plant |

**Table A2.2 Illustrative HAZOP study checklists**

| General List for Vessels | Vents/Drains |
|---|---|
| Pressure deviations | Isolation |
| Level deviations/bunding | Cross flow/reverse flow |
| Temperature deviations | Pressure rise/restriction/choke |
| Vortex | High back pressure in vent line |
| More/less mixing | Flashing/cooling |
| More/less layering | Luting |
| Increased velocity | Elevation change |
| Sediment | Sequence in the vent/drain header |
| Start-up/shutdown/purging | Air ingress |
| Fouling | Two-phase flow |
| Capacity | +Piping checklist |
| Dip legs | |

# An Illustration of HAZOP Study for a Continuous Operation

The model used for this illustration of the HAZOP study is of an off-shore wellhead gas platform linked to a central process facility (CPF) by a 15 km subsea pipeline. There is no chemistry but it is essential the team understands both the physics and physical chemistry of the process.

Gas is trapped in a loose sandstone formation 2000 m below the sea bed. It is in hydrostatic equilibrium, trapped by an impervious rock over and around the sandstone rock but in water contact at the bottom. The gas exists as a dense phase, mostly methane, saturated with water vapor at 80°C and 200 bar (20 MPa). The gas flows to the surface in a production tubing of 15 cm diameter, made up from a number of threaded sections, and the pressure falls due to both frictional losses and the reduced gas static head; the flowing pressure at the top of the well is about 125 bar. The production tubing is surrounded by a number of threading casings of increasing diameter which are used in the drilling program, the number and size of casings is a function of the local geology. The effective pressure containing capacity of each casing is a function of the strength of the rocks and the cement bond between the rock and tubing. If there is a leak of gas into the annulus between each casing, there is the potential for collapse of the inner casing due to pressure reversal, so it is essential to ensure a pressure gradient "in to out" and, if leakage occurs through the threaded sections of casing, it must be depressurized. Likewise, in the upper sections of the casing, multiple path leakage could lead to a fracture of the cement. There is one major barrier (flap-type valve) set in the production tubing 250 m below the sea bed. This is called the sub-surface safety valve (SSSV—sometimes called a down hole safety valve (DHSV)) and is held open by a hydraulic signal. Loss of the signal causes valve closure, and the valve is difficult to open under high pressure differential.

The casings are terminated on a "wellhead" (Figure A3.1, pages 108–109) which is bolted to the "Christmas tree." Within the tree is a master valve (MV) and, at an angle to the flow, a wing valve (WV); both are held open by a hydraulic signal. The emergency roles of each valve vary—the SSSV is protection against a main process event or failure of the tree, the WV is the main process valve, and the MV is used during downwell operations. Depending upon the level of emergency, the WV closes first, then the MV, and finally the SSSV. There are five wells in total in the field feeding the CPF. The flow of gas is controlled by a metal-to-metal seated manually operated valve called a choke. This is usually left in a fixed position and only adjusted occasionally. As the pressure drops across the choke, the temperature falls and two phases (condensate and gas) are produced. If the pressure drop is from 200 bar to less than about 50 bar, the temperature can fall below $0°C$ and ice and/or hydrocarbon hydrate solid can be formed which is controlled by injection of methanol. There is every potential for temperatures as low as $-50°C$ during the initial start-up of the process when the gas column in the production tubing loses its heat to the rocks surrounding the well and the initial temperature of the flowing gas could be as low as $15/20°C$.

The two phases flow into a collection manifold through a safety shut-off valve (ESDV2) into the subsea pipeline, about 30 m below sea level, linking the wellhead platform to the CPF, 25 m above sea level. The pipeline is rated for the maximum closed wellhead pressure (about 180 bar). There is some phase separation at low flow rates but for the most part transport is in mist or annular flow. The gas flows onto the CPF, 55 m above the sea bed, through a second safety shut-off valve (ESDV3) at the edge of the platform and then a process shutdown valve before entering a two-phase separator (Figure A3.2, pages 110–111) with a design pressure of 120 bar. Gas and liquid phases are metered separately, and the two phases then pass through a third safety shut-off valve (ESDV6) into a main subsea pipeline connecting the CPF to a shore terminal where it is processed. The data from the two flows, gas and liquid is used for reservoir performance monitoring and also apportioning products at the onshore terminal to each supplier. The main subsea pipeline has a pressure rating equal to the separator. A pig launcher can be fitted for pipeline monitoring (Figure A3.3, pages 112–113).

As the reservoir ages, the reservoir pressure falls, the flow rates decrease, and the water tends to increase. Ultimately, the reservoir tends to produce sand due to high-level pressure differentials and this is very abrasive.

Water associated with the gas is very saline and can be corrosive, so corrosion inhibitors are injected with the methanol used for hydrate suppression. The infield pipeline (and main pipeline) is protected from corrosion from seawater by sacrificial anodes, likewise the wellhead and CPF structures. There is therefore a potential electrochemical linkage between the three central elements. The pipe work is protected from corrosion internally by the corrosion inhibitor in the methanol injected for hydrate suppression.

The gas produces 1 $m^3$ of liquids per 20,000 standard $m^3$ gas during the separation process. The separator is designed for some "slugging" capacity and has some offline washing facility to remove sand. The liquid phase is level controlled and the gas phase is pressure controlled into the pipeline. The separator is fitted with high level alarms and shutdowns which close a shutdown valve inlet to the separation. The separator is protected against overpressure by a full-flow relief valve, discharging to a vent, sized for the maximum steady-state well flows. There are also two levels of pressure protection which first close the process shutdown valve and finally the WV.

There are other technical issues which are not discussed in this illustration as they do not serve to illustrate the study technique. These will need to be addressed in a real study.

## A3.1 METHANOL INJECTION

Methanol is used as a hydrate suppressant and is pumped by a positive displacement pump to injection pressure at the shore terminal. Corrosion inhibitor is mixed with methanol at the terminal and there is an emergency shut-off valve in the feed line at the wellhead platform (ESDV1). The main pump is fitted with a recycle pressure relief valve set at 240 bar. Each well is dosed continuously for 1 week with corrosion inhibitor. The changeover is carried out manually during the weekly inspection giving a 5-week cycle. The flow is controlled onshore to prevent hydrate formation in the pipeline.

## A3.2 GENERAL PROCESS DATA

In this model, the gas flow is taken as $10^5$ sm$^3$/h and the pipeline is 12 in. (30 cm) diameter. The closed-in system pressure is 180 bar for which all piping on the wellhead platform and the subsea piping is designed. The CPF is designed for a pressure of 120 bar downstream of the process shutdown valve (ESDV3).

All process piping is designed for $-30°C$ at the appropriate pressure, and the separator is designed for $-40°C$ as it will contain liquid hydrocarbon. During process blowdown, the gas temperatures can fall to $-40°C$ (or lower). Slugs and liquid equivalent to three riser lengths can be produced on increasing flow.

There is a fire and gas detection system which isolates the valves into and out of the CPF and depressure the process—level two. Such an event on this wellhead platform involves closure of the export valve and the WV plus MV but no depressuring. Loss of pressure in the manifold on the wellhead platform results in closure of the SSSV.

## A3.3 THE ISSUES

The inflow of gas is generally limited by the productivity index (PI) of the well. It self-limits at high demands and probably produces sand. Once the well is flowing, it must be managed to avoid sand (and water) production by fixing the choke position. Sand production can be detected by sand probes, and excessive sand production leads to erosion of the choke and piping and eventually settles out in the main pipeline. Corrosion is detected by a probe. More corrosion inhibitor is injected.

The SSSV is self-closing and cannot be opened with a pressure differential of more than a few hundred psi. SSSVs are available with compensating features and there are thousands of wells in UKCS. The MV and WV can be opened with a pressure differential and the choke operates with a pressure differential. The choke is not a shut-off valve and tends to wear and leak with time due to sand erosion.

Hydrates—a loose formation of hydrocarbon and liquid water—form above about 600 psi (4 MPa) and temperatures of about 15°C; this is controlled by methanol. Expansion of gas can produce both a water and hydrocarbon liquid phase. Temperatures as low as $-50°C$ are possible with throttling but are composition and pressure/temperature dependent.

When the infield pipeline is isolated, it slowly pressures due to choke valve leakage and could reach the shut-in well pressure. It is possible to open the shutdown valves with this pressure differential but valve seat damage may result.

The steady-state issues are generally sand and erosion; the dynamic, start-up, and shutdown issues are hydrants and low-temperature formation. The transient states involve the potential to move from wavy flow to mist flow with slug potential, which may be exacerbated by sea bed contours. The operating pressures are high and close to the piping design pressure limits.

There are effectively six blocks for this HAZOP study:

| 1) methanol pumps—onshore | not included |
|---|---|
| 2) wellhead platform | Figure A3.1 |
| 3) subsea infield line | Figures A3.1 and A3.2 |
| 4) production platform and vent system | Figures A3.2 and A3.3 |
| 5) main pipeline to the shore | Figure A3.3 (interface) |
| 6) onshore terminal | not included |

Block six—the onshore terminal has a slug catcher and gas processing equipment as well as the methanol pumps. These are out of the direct scope of this study, but it is very likely that this study will contain some actions for the hand over of information to the onshore HAZOP team.

## A3.4 METHODOLOGY

There are four distinct operations:

1. start-up—low-pressure downstream;
2. start-up—system pressurized;
3. shutdown and blowdown;
4. process transient.

There is little point in analyzing transients when the process cannot be started, so the logical approach is to analyze the start-up first (Tables A3.1 and A3.2, pages 114–120). Experience shows that many of the problems associated with the continuous processes occur during the dynamic phases of upset, start-up, and shutdown. The first part of the study illustrates this point and then in the second part, number 5 onward, moves on to the steady-state part of the study. It will be noted that the issues are quite different. It can be assumed that methanol is charged up to ESDV1, the process is air freed, and liquids are displaced so far as is possible prior to start-up.

| Team members | |
|---|---|
| Facilitator | Abe Baker |
| Project Manager | Charlene Doig |
| Platform Superintendent | Ed Fox |
| Process Engineer | Geoff Hughes |
| Instruments | Iain Joules |
| Scribe | Keith Learner |
| Petroleum Engineer | Mike November |

The design intent is to flow five gas wells at the combined rate of $10^5$ sm$^3$/h (85 mmscfd) of gas from the wellhead platform, with as low sand content as practicable, into a production separator on the CPF.

The team has the following available:

- a general description of the wellhead installation and the CPF;
- a selection of P&IDs;
- the "cause and effects" drawings for the shutdown system (Tables A3.3 and A3.4, page 121);
- the operating intent from which the detailed operations are written.

The outline operating intent is as follows:

- open SSSV using methanol to form a pressure balance;
- open MV and WV the choke valve and thereby pressure up the infield pipeline monitoring for evidence of chokes/hydrates;
- slowly pressure the separator and then also the main pipeline to the shore;
- each well is brought online in sequence.

The shutdown is on three levels.

**Level one—process upset**

Close an appropriate shutdown valve to arrest the cause of this event.

## Level two—major event

In general, this means a detected fire or detected gas leakage and closes all valves around the process and blowdown all vessels—pipelines remain pressurized.

## Level three—potential for major escalation

The SSSV on the wellhead platform can be closed by a manual signal from the control center on the central platform.

The riser ESD valves on the wellhead platform are closed by a manual signal from the control center on the central platform.

The riser ESD valves on the central platform are closed by fire or high-level gas detection local to the valve or by a manual signal from the control center.

### Failure mode

All valves are controlled by hydraulic power (not air) and all fail closed except for the blowdown valve ESDV7 which fails open.

### Piping code (for Figures A3.1–A3.3)

Fluid

G—gas
V—vent
M—methanol
D—drain
C—condensate
Pipe sizes are in inches
Pressure rating
1—ANSI class 150
9—ANSI class 900
15—ANSI class 1500
AP1 5000—special well piping design pressure 5000 psig
Materials
CS—carbon steel
SS—stainless steel

**See also the cause and effects tables (Tables A3.3 and A3.4).**

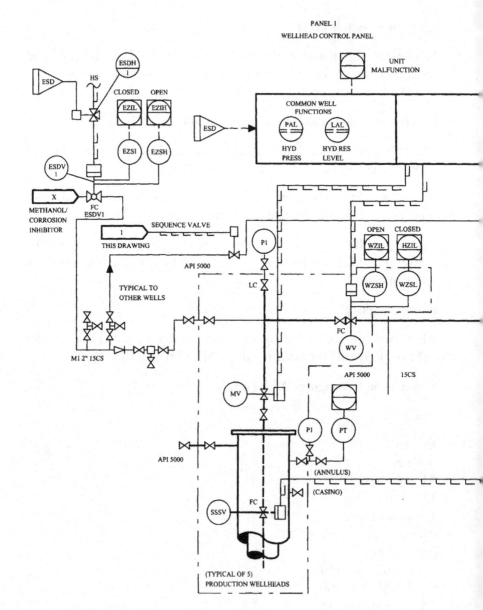

*Figure A3.1 P&ID 1 (to be used in conjunction with Table A3.1).*

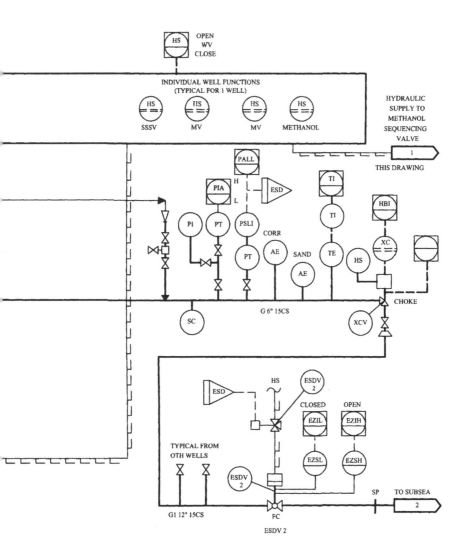

*Figure A3.1 (Continued).*

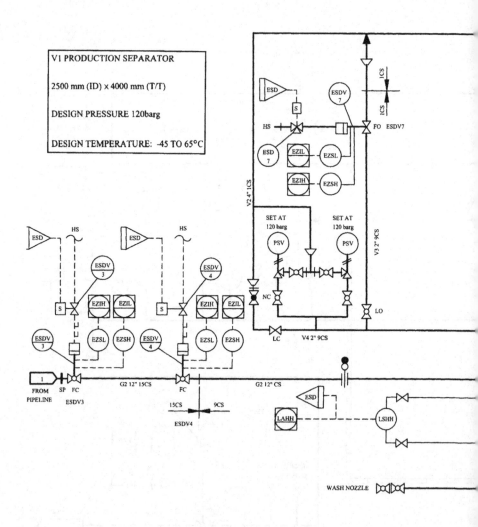

*Figure A3.2 P&1D 2 (to be used in conjunction with Table A3.2).*

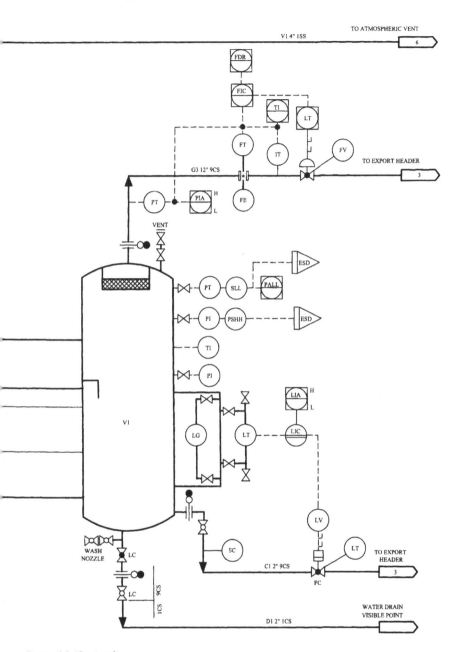

*Figure A3.2 (Continued).*

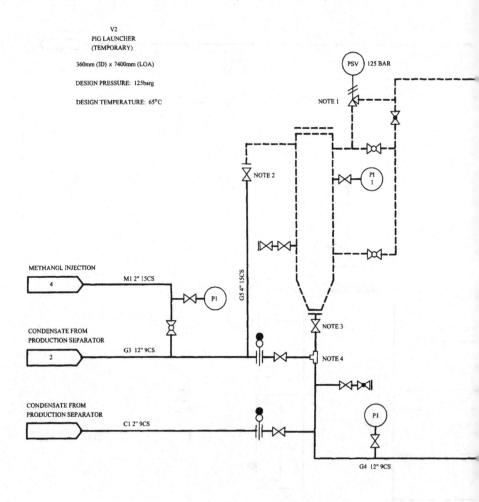

*Figure A3.3 P&ID 3 (to be used in conjunction with Table A3.2).*

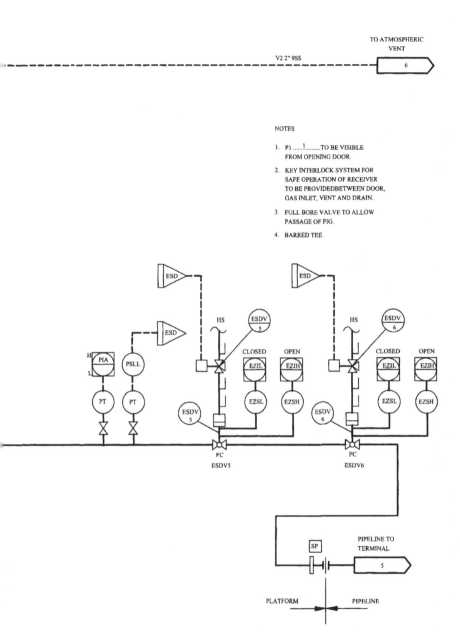

Figure A3.3 (Continued).

# Table A3.1 Demonstration of the HAZOP study conducted on node 1 (To be used in conjunction with figure A3.1)

DATE: 13/02/2015
NODE 1: Reservoir to Choke Valve Start-Up Operation
P&ID 1

INTENT 1: Pressure wellhead side of SSSV with methanol to allow SSSV to be opened
INTENT 2: To flow $2 \times 10^4$ sm$^3$/h of gas into the collection manifold G1-12″ 15 CS at a pressure of 100 bar and 15–20°C

STATUS: SSSV closed, MV closed, WV closed, choke closed, ESDV1 open, ESDV2 open, methanol pump running sequence valve open
ATTENDEES: AB, CD, EF, GH, IJ, KL, MN

| No. | Parameter | Guideword | Deviation | Cause | Effect | Protective Systems | Action | | Action on |
|---|---|---|---|---|---|---|---|---|---|
| 1 | Flow | No | SSSV closed | | | | | | |
| **LINE OUT THE METHANOL FROM THE INJECTION PUMPS THROUGH 2″ 15 CS TO EQUALIZE THE PRESSURE DIFFERENCE ACROSS THE SSSV** | | | | | | | | | |
| 2 | Pressure | Low | SSSV cannot be opened | • Pressure drop in M1 2″ 15CS is too high due to the need to inject methanol into online wells<br>• The pump capacity is inadequate | SSSV cannot be opened | RV on methanol pump is set to avoid overpressure of the methanol. Pump, pipeline and M1 2″ 15 CS (this pump is on shore) | 2.1 | Verify that the pressure drop in the subsea pipeline when dosing other wells is less than the valve set pressure minus 180 bar | CD |
| | | | | | | | 2.2 | Verify that the methanol pump has adequate capacity to dose other wells and pressurize the wellhead and down hole section of piping | CD |
| 3 | Pressure | Low | SSSV closed—no flow of methanol | Closed in system with a PD pump | Potential system overpressure | RV on pump will lift | 3.1 | Verify that the relief valve setting on the pump is correctly set to ensure all piping—on shore, offshore and subsea—is not over pressured | CD |
| | | | | | | | 3.2 | Ensure that the HAZOP of the methanol pump reflects that the pumps may run with a no flow case—consider the need for a pressure control spill valve round this methanol pump | CD |

**SSSV OPENED, MV OPENED, AND WV OPENED**

| No. | Parameter | Deviation | Causes | Deviation details | Consequences | Safeguards | No. | Actions | By |
|---|---|---|---|---|---|---|---|---|---|
| 4 | Pressure | Higher | Normal start-up | Wellhead pressure at closed-in condition | • Potential for reverse flow of gas to the shore if the methanol pump stops<br>• Pump does not inject methanol<br>• Potential reverse flow to production platform if NRV fails to open | • NRV fitted in methanol feed lines M1-2", 15 CS, M2-2" 15 CS<br>• PD pump is a form of NRV | 4.1 | Petroleum engineering to review maximum SITP and discuss with the project team | CD/MN |
| | | | Poor reservoir predictions | Shut in wellhead pressure is higher than anticipated | | | 4.2 | Verify the methanol pump relief valve is set at the correction pressure for both processes and piping | CD/MN |
| | | | | | | | 4.3 | Ensure the HAZOP of methanol pumps reflects the potential reverse flow through relief valve if fitted | CD |
| | Pressure<br>Temperature<br>Flow | Higher or lower – Higher Pressure discussed above | No meaningful deviations during the opening of the SSSV and pressuring to the choke | | | | 4.4 | Ensure the HAZOP of the methanol pumps reflects the hydraulic link from the well to the methanol pump with the potential for system over pressure if the suction isolation valve is closed | CD/EF |
| **NORMAL OPERATION** | | | | | | | | | |
| 5 | Flow | Lower | • Poor PI<br>• Hydrate/ice | Restriction in reservoir or downstream of choke | Loss of production | 1. None | 5.1 | Noted | |
| | | | | | | 2. Methanol | 5.2 | Ensure methanol injection rates are monitored and recorded daily at the shore | EF |
| 6 | Flow | Higher | Choke opened too far | | Possible sand production leading to erosion in piping and the choke | Sand probe (AE) | 6.1 | Ensure the peak flow characteristics are recorded in operating instructions | MN/EF |
| | | | | | | | 6.2 | Monitor sand probe on routine and more frequently early in the field life | LJ/EF |

| No. | Parameter | Deviation | Cause | Consequence | Safeguards | Rec. No. | Recommendation | Action |
|---|---|---|---|---|---|---|---|---|
| 7 | Flow | As well as | Sand production or well debris from drilling/perforation | • Erosion on piping or choke<br>• Possible choking of condensate control valve in V1 | Sand probe (AE) | 7.1 | Consider if a well cleanup program can be set in place | MN |
|  |  |  |  |  |  | 7.2 | Operating instructions should note the need to monitor for debris build-up in V1 | EF |
| 8 | Pressure | Lower | None |  |  |  |  |  |
| 9 | Pressure | Higher | Production higher than off-take<br><br>Production platform upset or shutdown and ESDV3 and 4 closed | • Pipeline pressurized to 180 bar<br>• On restart there is a high pressure drop over ESDV3 or 4 which may cause valve seat damage<br>• Initial gas flow through ESDV4 could overload the relief valves on V1 and over pressure V1 with a high transient flow | • Relief valves on V1<br>• Pipelines full pressure rated<br>• PSHH on V1 | 9.1 | Verify ESD3 has hard seats | IJ |
|  |  |  |  |  |  | 9.2 | Consider the need for a pressuring line around ESDV4 | EF/CD |
|  |  |  |  |  |  | 9.3 | Analyze the flow characteristics into V1 as ESDV4 is opened and the pressure/time profile in V1 | CD |
|  |  |  |  |  |  | 9.4 | Dependent upon 9.4 determine a means of establishing a steady dynamically limited flow which will not overpressure V1 | EF/GH |
| 10 | Pressure | Lower | Choke leaks and ESDV4 leaks plus platform blow down plus WV closed | • Lower temperatures<br>• scenario unlikely | None | 10.1 | Verify there is no thermal implication in the choke | MN |
|  |  |  |  |  |  | 10.2 | Review this scenario with respect to the pipeline later in the study | AB |

**See 10.1 and 10.2**

| No. | | | | | | | Ref | Action | Initials |
|---|---|---|---|---|---|---|---|---|---|
| 11 | Temperature | Lower | | | | | See 10.1 and 10.2 | | |
| 12 | Maintenance | None | Can the items up to and including the choke be maintained | Poor isolation standards | Loss of production | Isolation valve | 12.1 | Can the sand probe and corrosion probe be removed safely; are they fitted in self-isolation pockets? | IJ/CD |
| | | | | | | | 12.2 | Review the need for double isolation on each well at the manifold | MN/CD |
| | | | | | | | 12.3 | Determine if wear on the choke is likely to be significant at any phase of the field life | MN/CD |

*It will be noted that, sometimes, there are two persons in the "actions on" part of the table. This is because these two were the leaders of the discussion and are the most likely to understand the issues. The first person (initials underlined) is the one who is accountable for the action.*

*Please note: " = inches.*

*No other meaningful parameters and deviations were found and the study of section/mode was completed.*

*Others to be analyzed:*
- *corrosion;*
- *erosion;*
- *access for maintenance;*
- *maintenance features—standards of isolation;*
- *purging features—vents and drains, location and termination points;*
- *diagnostic features.*

## Table A3.2 Demonstration of the HAZOP study conducted on node 2 (To be used in conjunction with figures A3.2 and A3.3)

*DATE:* 13/02/2015
*NODE 1:* Subsea Pipeline from Choke to ESDV3 Start-Up/ Operation
*P&ID 2 and 3*

*INTENT 1:* To flow $10^5$ sm$^3$ of gas to the production platform at a pressure of 100 bar and 15–20°C
*STATUS:* All SSSV open, all MV, WV open, ESDV3 and 4 open and chokes closed.
Initial state 0 bar, nitrogen filled
*ATTENDEES:* AB, CD, EF, GH, IJ, KL, MN

| No. | Parameter | Guideword | Deviation | Cause | Effect | Protective Systems | Action | | Action on |
|---|---|---|---|---|---|---|---|---|---|
| **Slowly open choke** | | | | | | | | | |
| 13 | Pressure | Low/lower | Pipeline pressure low | Pressuring | Low-temperature ice or hydrate formation | Methanol injection | 13.1 | Review the temperature/time profile as the pipeline is pressured taking into account the thermal mass of the pipework—the lowest temperature will be at the choke | CD |
| | | | | | | | 13.2 | Pursue means of pressuring the system from the onshore terminal | GH, CD, EF̲ |
| 14 | Pressure | High | | | | Pipeline fully pressure rated | | Noted | |
| 15 | Temperature | Low | See 13 | See 13 | See 13 | See 13 | | See 13. Consider again under higher temperature | AB |
| 16 | Temperature | High | Adiabatic compression of nitrogen | Nitrogen piston compressed by incoming gas | None | None | 16 | Noted. Review means of displacing nitrogen—a potential contamination in gas as part of 13.2 | GH, CD, EF̲ |

| No. | Parameter | Guide Word | Cause | | Consequence | Safeguard | Ref. | Action | Code |
|---|---|---|---|---|---|---|---|---|---|
| 17 | Flow | Low/no/high | No logical meaning during start-up | | | | | | |
| 18 | Phase | Change | Production of ice, hydrate, condensate | Expansion of gas into the pipeline | Potential choke | Methanol injection | | Noted | |
| | | | | | | | 18.1 | Ensure the operating instructions record the need for continuous methanol upstream of the choke while pressuring the line | EF |

**NORMAL OPERATION**

| No. | Parameter | Guide Word | Cause | | Consequence | Safeguard | Ref. | Action | Code |
|---|---|---|---|---|---|---|---|---|---|
| 20 | Flow | Low | Low production | Low off take at terminal | Potential slugging regime | None | 20.1 | Review line slug size and separation/hold up in capacity in V1 | GH |
| 21 | Flow | Higher | Rate increase | Higher off take at the terminal | Potential slug formation and reactions forces on the riser | | 21.1 | Include in 20.1 | GH |
| | | | | | | | 21.2 | Review the riser support against slugging | CD |
| 22 | Flow | High | High demands | | Possible sand formation and erosion | Sand probes in each well | 22.1 | See 6.1 and 6.2 | |
| 23 | Flow | Lower | Restricted flows of the pipeline | Hydrate formation | Line choked ice/ hydrate slug may move causing reactions forces on the riser and sudden high flow into V1 | | 23.1 | Devise means of avoiding hydrate plugs moving during recovery from a hydrant plug | EF |
| | | | | | | | 23.2 | Monitor methanol injection daily on shore | EF |

| | | | | | | Methanol | | | |
|---|---|---|---|---|---|---|---|---|---|
| 24 | Flow | High | High flow | Hydrate slug moves when under high pressure differential | Higher pressure in V1 | | 24.1 | See 21.2 | EF/GH |
| | | | | | | | 24.2 | See 9.4 and 9.5 | |
| 25 | Temperature | Lower | | Pipeline depressured | Possible hydrate formation | | 25.1 | Review the temperature in the pipeline during depressuring — verify if it does not go out of the spec limits. Allowance should be made of heat flow into the line. See 23.1 and 24.1 | GH |
| 26 | Temperature | Higher | Pipe warmer than when laid | Hotter fluids flowing in pipeline after start-up | Thermal expansion of the pipeline | | 26.1 | Consider the potential for upheaval buckling and the need for trenching or rock dump | EF/GH |
| 27 | Electro potential | High differential | Possible loss to cathodic protection | Localize corrosion outside the pipeline | • Sacrificial anodes<br>• Isolation flanges | | 27 | Routinely monitor the performance of the insulating flanges at the wellhead and production platform | EF/IJ |
| 28 | To be continued | | | | | | | | |

*It will be noted that, sometimes, there are two persons in the "actions on" part of the table. This is because these two were the leaders of the discussion and are the most likely to understand the issues. The first person (initials underlined) is the one who is accountable for the action.*

Other issues are:

- corrosion internally and externally on the process piping and subsea pipelines. The process and pipelines are electrically insulated by a special flange arrangement;
- erosion;
- nitrogen disposal at start-up.

**Table A3.3 Cause and effects for wellhead platform**

|  | Detected Gas (Low Level) | Detected Gas High Level 60% LEL | Detection Fire | Vibration (Impact) |
|---|---|---|---|---|
| WVs | C | C | C | C |
| UM valves |  | C | C | C |
| SSS valves |  |  | C | C |
| ESDV1 |  |  | C | C |
| ESD2 |  |  |  | C |
| C—Closed |  |  |  |  |
| O—Open |  |  |  |  |

**Table A3.4 Cause and effects for CPF**

|  | Local Fire or Gas Detected at ESDV3 and 6 | V1 High Pressure | V1 High Level | General Gas Detection High Level 60% | General Fire |
|---|---|---|---|---|---|
| ESDV3 | C |  |  |  |  |
| ESDV6 | C |  |  |  |  |
| SD wells | C |  |  |  |  |
| WV |  | C | C | C | C |
| ESDV4 |  | C | C | C | C |
| ESDV5 |  | C | C | C | C |
| ESDV7 |  | C | C | O | O |
| C—Closed |  |  |  |  |  |
| O—Open |  |  |  |  |  |

# An Illustration of HAZOP Study for a Batch Operation

## A4.1 INTRODUCTION

This example illustrates the following aspects of a HAZOP study:

- an overall description of the plant and process;
- the selection of the stages of the process for study;
- a list of the documentation available to the team;
- the relevant P&ID;
- the team membership;
- for one stage:
  - the detailed description and the design intention;
  - a list of the parameters and guidewords used;
  - a part of the HAZOP study report, illustrating some of the deviations with consequences, safeguards, and actions.

The example has been fabricated, although it is intended to resemble a real operation. For simplicity, the main reactants are simply labeled A and B.

## A4.2 THE COMPANY, SITE, PLANT, AND PROCESS

The company is a long-established general chemical manufacturing company, employing over 2000 people at several sites in the UK. It has a good safety record and routinely uses HAZOP study for new plant, new processes, and major modifications. A central safety group oversees this activity and provides trained leaders for all process hazard studies on new plant or processes. The six-stage process hazard study system is used, with HAZOP study as the usual method used at stage three.

The site concerned with this process is on the outskirts of an industrial city and employs 350 people. It lies between a river and a major road and, on one side, is close to an old housing estate. There are a number of continuous processes and several general-purpose batch units on-site. The operation to be examined is for a new process in one of the batch units which will be adapted to the needs of this process. Laboratory work has been done to determine the batch size and conditions. Reaction hazard investigations have been carried out to identify the reaction hazards and to define a basis for safety.

The essential elements of the process are as follows. The sodium salt of an organic reactant, A, is formed by adding caustic soda solution to A in a large reaction vessel. The process is mildly exothermic; a slight excess of caustic is used to maintain the pH around 11. The salt is then reacted with a second organic material, B, added at a controlled rate from a measure vessel. This reaction is very exothermic, and cooling is required. To obtain a high-quality product, the reaction needs to be carried out at between 55−60°C.

The section of the batch plant to be used for the reaction stage consists of two measure vessels at level two that will be used for the caustic soda solution and for reactant B. The required amounts are taken from drums on weigh scales, using vacuum. The feed to the reactor at level one is by gravity. Component A is pumped from drums directly into the reactor followed by a line flush with water. After the reaction is complete, as confirmed by analysis of a sample, the products are pumped to another vessel for further processing.

Component A is a solid, mp 30°C, flash point 45°C, with a long-term exposure limit (LTEL) of 10 ppm. Component B is a liquid, mp 5°C, bp 122°C, flash point 10°C, with a LTEL of 2 ppm and it causes chemical burns on skin contact. The solution of A in caustic soda and the product solution after reaction of B are all single-phase systems.

## A4.3 THE PROCESS STAGES SELECTED FOR HAZOP STUDY

From consideration of the draft operating procedure, based on the laboratory investigations, the stages marked ✓ were selected for HAZOP study:

|   | 1 | Check plant set-up |
|---|---|---|
| ✓ | 2 | Measure 250 kg of 30% caustic solution to vessel F1 |
|   | 3 | Melt two drums of A in a drum heater |
| ✓ | 4 | Pump 425 kg of A to reaction vessel F3 |
|   | 5 | Flush line with 100 L of water |
|   | 6 | Heat F3 and contents to 55°C |
| ✓ | 7 | Run caustic from F1 to F3 |

| ✓ | 8 | Measure 375 kg of component B to vessel F2 |
|---|---|---|
| ✓ | 9 | Reaction stage: controlled feed of B from F2 to F3 |
| ✓ | 10 | Workout for 20 min (combined with step 9 for HAZOP study) (see Table A4.1, pages 132–137) |
| | 11 | Sample and check product |
| ✓ | 12 | Pump contents of F3 to F4 |
| | 13 | Wash F3 with 200 L of water |
| | 14 | Pump wash water to F4 |

The basis for not selecting some stages is that they are simple, familiar steps of low hazard potential, with little chance of incorrect execution or omission by the operators.

During the HAZOP study, the team is expected to consider all potential SHE hazards as well as operability problems.

## A4.4 HAZOP STUDY OF STEPS 9 AND 10, THE REACTION STAGE AND WORKOUT

The HAZOP study team

| Name | Discipline | Job Title | Role/Represents | Years |
|---|---|---|---|---|
| Mike Stopner | Chemist | Safety Advisor | Leader | 25 |
| Jennie Howard | Chem Eng | Project Engineer | Design team | 7 |
| Tom Bailey | | Shift supervisor | Operations | 17 |
| Bob Teryl | Chemist | R&D Chemist | Process development | 12 |
| Andy Wires | Electrical Eng | Control Engineer | Control/instruments | 3 |
| Frank Laycrew | Mech Eng | Site Engineer | Services/maintenance | 14 |
| Karl Jones | Chem Eng | Trainee Project Engineer | Scribe | 1 |

All are full-time employees of the company and, apart from KJ, have previous experience of HAZOP study, including training on a 2-day in-house course. MS has also attended an external 4-day training course on HAZOP study leadership and has been leading studies throughout his time in the central safety department (5 years).

## A4.4.1 Documentation for the Study

Documents include a set of P&IDs showing the plant as it will be set up for the operation including spaded lines and any new connections. A separate P&ID is used for each stage of the HAZOP, marked up showing the actual items of the plant involved in that stage and

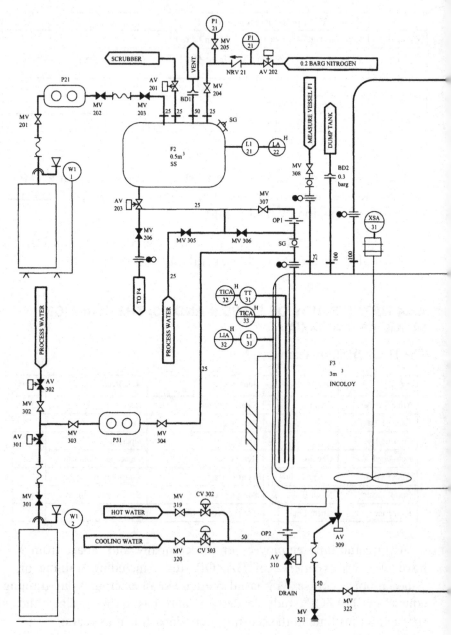

*Figure A4.1 P&ID for steps 9 and 10; drawing on AB01/Rev 2 (to be used in conjunction with Table A4.1, pages 132–137).*

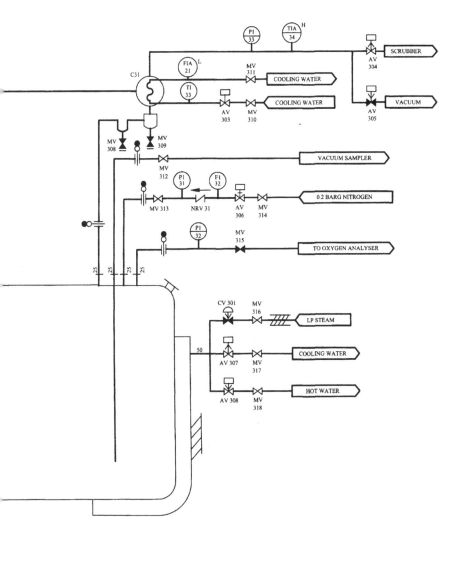

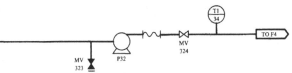

*Figure A4.1 (Continued).*

showing which valves are open (Figure A4.1 for steps 9 and 10). Full details of plant and equipment specifications are available if required. The other items made available to the team are:

- site plan;
- process description and outline operating procedure;
- reaction hazard review covering laboratory studies of the chemistry, reaction kinetics and thermodynamics. Results from differential scanning calorimetry analysis of components A, B, and the reaction product; adiabatic calorimetry data, including reaction simulation. The basis of safety was derived from this work;
- material safety data sheets for all reactants; the available hazard data for the product;
- alarm, interlock and trip schedule;
- reports from Hazard Studies 1 and 2.

## A4.4.2 Plant Conditions and Step Description

As a consequence of the previous steps, the state of the plant at the start of this step is that the reaction vessel, F3, contains 425 kg of component A, 100 L of water, and 250 kg of 30% caustic. Component A has been converted to the sodium salt and sufficient excess caustic is present to raise the pH to 11. The stirrer is running and the mixture is controlled at 55–60°C, using hot water to the vessel jacket. The vessel is open to the scrubber through a condenser, set to return any condensed liquids to F3. Measure vessel F2 contains 375 kg of component B that is to be run down to F3 over a period of about 3 hour at 2 kg min$^{-1}$. The flow rate is controlled by the orifice plate, OP1. Both F2 and F3 are operated at atmospheric pressure under nitrogen to prevent the formation of a flammable atmosphere. The continuous feed of nitrogen into each vessel is vented through the scrubber. The scrubber is operated to reduce the levels of vapor A and B to below their LTEL values to permit safe discharge at a high level.

The basis for safety was developed from work initiated in HS 2. It requires:

- The rate of heat evolution from the amount of B present in the reactor does not exceed the cooling capacity of the reactor. For this to be achieved, all the following conditions must be maintained:
  - the rate of addition of B is controlled at 2 kg min$^{-1}$;
  - there is continuous stirring to avoid accumulation of B;
  - the temperature is kept above 40°C to minimize accumulation of B;

- If runaway reaction does occur, the reaction vessel is protected by the rupture of the bursting disc which relieves to a dump tank from which the resulting vent is acceptable at the expected frequency.

From the earlier hazard reviews, the worst case event is identified as a runaway reaction which is not fully relieved by the bursting disc. In this circumstance, it is estimated that the reactor design pressure could be exceeded by a factor of 1.5. Reactor rupture is possible but is of low probability. The consequences of vessel rupture would be the possibility of operator fatality and severe local contamination which, if an aerosol cloud forms, could be blown off-site. The worst effect of this is if the wind direction was toward the local housing estate. The operation of the bursting disc is crucial to avoidance of this event, and it has been sized on the basis of small-scale experiments providing data for use with the DIERS design rules.

As the previous steps have been subjected to HAZOP study, it is assumed that there is negligible chance that they are not as described unless new causes are found.

Step nine is started by opening valve AV203, all other conditions having been set previously. The operator initiates the step from a control panel through the computer control system. Apart from occasional checks during the 3 hour addition, the operator relies on the alarms to indicate any deviation from the set conditions. Alarms are set to indicate stirrer motor failure, and low and high temperatures of 50°C and 65°C, respectively. If the temperature reaches 70°C, the valve AV203 is automatically tripped to close. Step 10 (working out for 20 min) follows on directly from step nine with no change in the system or the settings. At 3½ hour from the start of the reaction stage the computer closes AV203, puts the system to hold and awaits an operator input.

## A4.4.3 Parameter and Guidewords
The following guidewords (and abbreviations) are used.
N    No (not, none)
M    More (more of/higher)
L    Less (less of/lower)
R    Reverse
PO    Part of

AWA    As well as (more than)
WE     Where else
EL     Early/late
O      Other (other than)

The team leader has prepared preselected combinations of parameters and guidewords that give meaningful deviations, as shown in Table A4.2 (page 138). Team members are given opportunities to extend this list. Only those combinations generating a significant discussion are recorded in the HAZOP tables as no realistic meanings or likely causes were found for some. It was found that some of the later combinations had been adequately examined under earlier pairings—for example, "part of composition" was mostly dealt with by "more/less quantity." Some additional deviations came up during the analysis.

The following "parameters" are also considered under the guideword "other (other than)":

- services (including failure modes of valves);
- maintenance;
- safety;
- process interruption/hold/recovery;
- drainage;
- trips;
- corrosion;
- ignition sources (e.g., static electricity).

Action in emergencies—for example, fire, explosion, and toxic leak—is considered for the whole process at the completion of the HAZOP study of all the individual stages.

## A4.5 A SECTION OF THE HAZOP STUDY REPORT FOR THE BATCH REACTION (SEE TABLE A4.1)

Step—reaction and workout, steps 9 and 10 (see Table A4.1, pages 132–137);
Team—MS (leader), JH, TB, BT, AW, FL, KJ (scribe);
Drawing—P&ID (see Figure A4.1, pages 126–127);
Meeting date: 28/02/15; Revision: 0.

## A4.5.1 Step Description
A measured quantity of component B is added by gravity feed from F2 to the prepared mixture in F3 at 55–60°C. The step is initiated by the operator using the computer control system and the addition is started by opening valve AV203, all other valves being preset. The addition takes about 3 hours followed by a short workout period. The total time for these steps is 3½ hour after which AV203 is closed and the system held awaiting an operator command.

## A4.5.2 Design Intention
To transfer by gravity from the measure vessel F2, and to completely react, 375 kg of component B with the stirred aqueous solution of the sodium salt of 425 kg of component A in F3. The transfer rate is to be controlled at $2 \, \text{kg min}^{-1}$ by orifice plate OP1. The reaction temperature in F3 is to be controlled in the range 55–60°C. On completion of the addition, the reaction mixture is stirred for 20–30 min before sampling. A nitrogen atmosphere is maintained in F2 and F3 at the flow rates established in previous steps.

### Note 1
This report is from the first analysis by the team. Since the earlier steps in the process have already been studied, some deviations—for example, wrong amount of A is present—have already been considered. Entries will only occur for these deviations for new causes or new consequences suggested by the team. The team took one session developing this report.

### Note 2
The numbering system adopted is to have an item number for every row of the analysis and to relate the actions to that number. Where two or more actions result they are numbered as, for example, 14.1 and 14.2. Some action numbers are not used such as 15 and 16.

### Note 3
A response/comment column is available but not shown. It is used to enter the responses to the actions and to record any further comments by the HAZOP study team.

**Table A4.1 HAZOP study on steps 9 and 10 (reaction and workout) (to be used in conjunction with figure A4.1, pages 126–127)**

| Ref. | Parameter | Deviation | Possible Cause | Consequence | Safeguard/Protection | No | Action | On |
|---|---|---|---|---|---|---|---|---|
| 1 | Quantity/step | No B is added — step omitted | Operator error, for example, at shift handover. MV307 closed after maintenance | Spoilt batch | Detected at sampling and can easily be corrected. | 1.1 | Start-up check to confirm that MV307 is open | TB |
| | | | | | Batch sheet requires analysis to be signed off by supervisor | 1.2 | Operating procedure to include a sight glass check that flow is established | TB |
| 2 | Quantity | Excess of B is added | F2 not fully emptied from last batch | Excess of B in product: batch will be out of specification | Detected at sampling but a special procedure will then be required | 2.1 | Operating procedure to include a check on vessel F2 before B is measured out | TB |
| 3 | Quantity | Too little of B is added | Blockage in line or at OP1 | Batch out of specification and process delay | Detected at sampling | 3.1 | Check procedure for clearing line and OP1 when transfer line holds component B | FL |
| | | | | | | 3.2 | Batch sheet to require a check that F2 is empty at end of the addition stage | TB |
| 4 | Quantity | Too much of A is present | Error at earlier stage resulting in small excess (double charging covered in HAZOP of addition step) | Batch out of specification and process delay | Detected at sampling | 4.1 | Check that procedure will be written to cover this case and include in training program | JH |
| 5 | Quantity | Too little A is present | Error at earlier stage resulting in small deficiency | Batch out of specification and process delay. Not easily corrected | Detected at sampling | 5.1 | Evaluate likelihood of this deviation and, if necessary, draw up procedure | MS |
| 6 | Flow (rate) | Too fast | Corrosion/erosion of OP1 | Reaction rate and heat release increased. May eventually exceed vessel cooling capacity leading to over-temperature | Independent alarms TICA 32/33 located in a manned control room | 6.1 | Check that OP1 material is compatible with component B | BT |

| | | | | | | | |
|---|---|---|---|---|---|---|---|
| 7 | Flow (rate) | Too fast | Wrong OP fitted at OP1 after maintenance | Could quickly exceed the vessel cooling capacity, causing a reaction runaway and demand on BD2 | TICA32/33 located in a manned control and BD2 relieving to dump tank. Good control of maintenance | 7.1 | Specify OP1 size in operating procedure and ensure problem is covered in operator training | TB |
| | | | | | | 7.2 | Confirm flowrate at OP1 at the water trials stage | FL |
| | | | | | | 7.3 | Control sequence to include trip closure of AV203 and fully open CV303 in the event of over-temperature | AW |
| 8 | Flow (rate) | Too fast | MV306 is open and so orifice plate OP1 is bypassed | Will very quickly exceed the vessel cooling capacity and lead to a reaction runaway and demand on BD2 | TICA32/33 to manned control room and BD2 relief to dump tank. BD2 is sized for addition at maximum possible flow rate in a 25 mm line | 8.1 | MV306 to be locked closed as it is not used in this process | FL |
| | | | | | | 8.2 | Include sensing of BD action to give alarm and to close AV203 | AW |
| | | | | | | 8.3 | Consider removal of OP2 from the cooling water inlet line so full cooling capacity will be available. Take into account the original purpose of OP2 in controlling heating rates/cooling profiles/blowdown of condensate | FL |
| 9 | Flow (rate) | Too slow | Partial blockage in line or at orifice plate OP1 | Batch time extended | Operator will note problem when seeking to move to next stage | 9.1 | Covered by actions 3.1 and 3.2 | FL TB |
| 10 | Flow | Elsewhere | Crack or leak at BD2 (action 8.2 only detects full burst) | Loss of contaminated nitrogen to dump tank and eventually to atmosphere | None | 10.1 | Put BD2 on a regular checking schedule | FL |

(Continued)

## Table A4.1 (Continued)

| Ref. | Parameter | Deviation | Possible Cause | Consequence | Safeguard/Protection | No | Action | On |
|---|---|---|---|---|---|---|---|---|
| 11 | Temperature | High | Control problem or faulty temperature signal (reads low) | Overheating will occur, with contributions from the heating system. Most serious condition would be common effect since both temperature probes are in the same pocket in F2 | None unless the fault also leads to a low temperature alarm when operator intervention could be expected | 11.1 | Check whether it is possible to physically separate the two temperature probes (control and protection) to reduce common cause effects | FL |
| 12 | Temperature | High | Loss of cooling water (a low probability event) | Overheating. Runaway if cooling water is not restored or the addition halted | TICA32/33 are located in the manned control room and BD2 relieves to dump tank | 12.1 | Covered by action 7.3 | AW |
| 13 | Temperature | High | Jacket not switched from steam to cooling water after earlier step | Overheating with possible reaction runaway | TICA32/33 to manned control room and BD2 relieves to dump tank | 13.1 | Control program to include checks that valve CV301 on the steam line is closed | AW |
| 14 | Temperature | Low | Control problem or faulty temperature signal (reads high) | Poor quality batch. Extreme outcome is cessation of reaction and accumulation of unreacted B | TAL from TICA 32 | 14.1 | Take TAL from both the control and the protection temperature sensors | AW |
| | | | | | | 14.2 | Determine suitable interval for calibration checks on TICs | FL |
| 15 | Pressure | High/low | No causes identified in addition to the runaway situations discussed above | | | | | |
| 16 | Reaction rate | High/low | No additional causes found | | | | | |

| No. | Parameter | Guide word | Cause | Consequence | Safeguards | Action ref. | Action | By |
|---|---|---|---|---|---|---|---|---|
| 17 | Mix | No mixing | Mechanical coupling fails or agitator blade becomes detached | Risk of accumulation of unmixed B leading to uncontrolled reaction | Possibly detected by low motor current alarm | 17.1 | Add a rotation sensor to the shaft of the stirrer; interlock to reactant feed valve AV203 | AW |
| 18 | Mix | No mixing | Motor failure | Risk of accumulation of unmixed B leading to uncontrolled reaction | Alarm on motor current (low) | 18.1 | Existing safeguard adequate provided action 17.1 is implemented | AW |
| | | | | | | 18.2 | Develop a safe operating procedure for restarting a batch after accumulation has occurred | BT |
| 19 | Mix | Less mixing | Viscous mixture formed | Stirring becomes inefficient and unmixed B may accumulate | May be alarmed by sensor added in action 17.1 | 19.1 | Check viscosity under extreme conditions to decide if action is needed. If so, include an alarm on high motor current | BT |
| 20 | Mix | Reverse | Incorrect connection after maintenance | Stirring becomes inefficient and unmixed B may accumulate | None | 20.1 | Include a check on stirrer operation in the commissioning trials and in the maintenance procedures | TB |
| 21 | Composition | Part of | Wrong ratio of reactants covered under high/low quantity | | | | | |
| 22 | Composition | As well as | Wrong drum used when charging component B | Unpredictable but minimum will be a spoilt batch | Covered in HAZOP of the charging step | 22.1 | Review actions from earlier HAZOP and ensure that the purchasing department specifies a distinct drum color | MS |
| 23 | Control | None | Complete loss of control computer | System moves to fail safe condition | Design assumes a period of operation of the computer on its UPS. Ultimate protection is provided by BD2 | 23.1 | Check that fail safe settings include isolation of feed of B, continued stirring and full cooling to vessel jacket | AW |

(Continued)

## Table A4.1 (Continued)

| Ref. | Parameter | Deviation | Possible Cause | Consequence | Safeguard/Protection | No | Action | On |
|---|---|---|---|---|---|---|---|---|
| 24 | Control | Part of | Selective failure. Most serious would be loss of temperature sensors/control | Possible undetected overheating | Ultimate protection is provided by BD2 | 24.1 | Check that the temperature sensors connect to different input boards | AW |
| | | | | | | 24.2 | Include temperature comparison (TICA32/22) in the checks and add a difference alarm | AW |
| 25 | Operator action | Sooner | Step started early | Starting temperature is low. Reactant may accumulate and then cause runaway reaction once mixing starts | Ultimate protection is provided by BD2 | 25.1 | Specify the lowest safe starting temperature | BT |
| | | | | | | 25.2 | Provide software interlock to prevent low temperature start | AW |
| 26 | Operator action | Part of | Workout period is shortened if the addition is slow (for any reason) | Uncertain—basis for inclusion of the workout period is not clear | | 26.1 | Carry out further laboratory work to determine the importance of the workout and to define the minimum allowable time | BT |
| 27 | Services | Loss of instrument air | | All valves move to assigned failure positions | | 27.1 | Review the failure modes of all valves to ensure specification is correct | JH |
| 28 | Services | Power loss | Unpredicted failure, cut cable, and so on | Stirrer stops. Computer moves plant to a safe hold position | Computer has its own UPS | 28.1 | Include this condition in the check under 27.1 | AW |
| | | | | | | 28.2 | Consider need for planned restart procedure after such an interruption | JH |
| 29 | Maintenance | Work on AV203 | Valve problem on AV203 during the transfer | AV203 cannot be isolated from F2 for safe maintenance | None | 29.1 | Put additional manual valves in the F2/F3 line | FL |
| | | | | | | 29.2 | As a general action, review the P&ID to ensure all key items can be isolated | FL |

| | | | | | | | | |
|---|---|---|---|---|---|---|---|---|
| 30 | Vessel entry (F3) | Other activity | Inspection or other requirement for entry to vessel | Risk to operator from inert atmosphere, especially nitrogen | Spades installed on all lines | 30.1 | Review the isolation of F3, including possible insertion of flexible section into the nitrogen line so that it can be disconnected and blanked off. Need to cover F2 as well since it has its own nitrogen supply and is linked to F3 | MS |
| 31 | Drainage | Leak of B | Leaking flange on transfer line from F2 to F3 | Some loss of component B into process area | All spillages in this area run to a common sump | 31.1 | Check the materials in use on adjacent units for potential incompatibility | FL |
| | | | | | | 31.2 | Operating procedure to include a routine inspection of the transfer line at the stage of the process | TB |
| 32 | pH | High/low | Imbalance in quantities of A or caustic added previously | Batch quality affected unless initial pH is range 10–11.5 | None | 32.1 | Operating procedure to include a check on pH before this step is initiated | TB |
| | | | | | | 32.2 | Consider need for a procedure for correction of pH | BT |
| 33 | Trip action | Out of range condition | Any | Control system moves the plant to a predetermined state based on the trip signals | | 33.1 | Prepare matrix to show which valves act in each trip scenario. Review the matrix at next HAZOP meeting | JH |
| 34 | Operator PPE | Exposure | Leakage or spillage | Contamination | Standard procedures | 34.1 | Confirm that procedures exist for all materials handled in the process | TB |

| Table A4.2 Preliminary list of applicable combinations of parameters and guidewords | | | | | | | | | |
|---|---|---|---|---|---|---|---|---|---|
| **Guidewords** | | | | | | | | | |
| **Parameter** | **N** | **M** | **L** | **R** | **PO** | **AWA** | **WE** | **EL** | **O** |
| Quantity | ✓ | ✓ | ✓ | | ✓ | | | | |
| Flow | ✓ | ✓ | ✓ | ✓ | | | ✓ | | |
| Temperature | | ✓ | ✓ | | | | | | |
| Pressure | | ✓ | ✓ | | | | | | |
| Reaction | ✓ | ✓ | ✓ | ✓ | ✓ | ✓ | | | |
| Mix | ✓ | | ✓ | ✓ | | | | | |
| Step | ✓ | | | | | | | ✓ | |
| Control | ✓ | | | | | ✓ | | | ✓ |
| Composition | | | | | ✓ | ✓ | | | |
| Operator action | ✓ | | | | | ✓ | | ✓ | ✓ |

# An Illustration of HAZOP Study for a Procedure

This study is loosely modeled on an article in the *ICI Safety Newsletter* No 32 August 1971. As the study is "hypothetical," the working parameters of the up- and downstream processes are not available but this should not detract from demonstrating the study process. Also, as the study is clearly short and operations oriented, it does not justify a full team and the Facilitator may also act as Scribe. With the limited number of team members, some of the actions must be dealt with by someone outside of the study group who has the skills so to do. It is the responsibility of the person named in the study records to ensure that a competent person answers them and that they are implemented properly.

## A5.1 BACKGROUND

This example is based upon the HAZOP study of a planned modification of an existing process operation (Figure A5.1).

An intermediate storage tank (IST) receives a $C_6$ hydrocarbon stream (averaging 25 m$^3$/hour) from the reflux drum of an atmospheric pressure distillation column, run down on exit level control via the reflux pumps into the 250 m$^3$, nitrogen-blanketed tank. This conical-roofed tank serves as a buffer and temporary storage for the material before the $C_6$ material is pumped by the J1 centrifugal pump, on level control, to the plant petrol blending unit. The IST operates at ambient temperature and at 500 Pa on split range pressure control and is inerted by nitrogen from the 1.3 bar site nitrogen supply. The tank is protected by a pressure (vacuum) valve (PV) set at $-250/+750$ Pa. It is in a bunded enclosure with an overflow, sealed with glycol, which empties into the bund. There is adequate instrumentation, including level indication with high- and low-level alarms and high-level trip plus temperature and pressure indication, all to the site control room.

# Intermediate storage tank and link to the petrol blending system

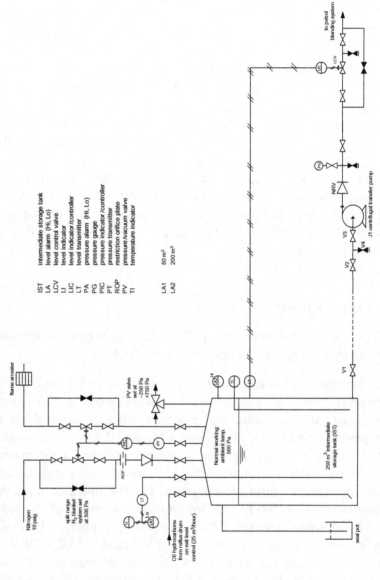

| | |
|---|---|
| IST | intermediate storage tank |
| LA | level alarm (Hi, Lo) |
| LCV | level control valve |
| LI | level indicator |
| LIC | level indicator/controller |
| LT | level transmitter |
| PA | pressure alarm (Hi, Lo) |
| PG | pressure gauge |
| PIC | pressure indicator/controller |
| PT | pressure transmitter |
| ROP | restriction orifice plate |
| PV | pressure/vacuum valve |
| TI | temperature indicator |

| | |
|---|---|
| LA1 | 50 m³ |
| LA2 | 200 m³ |

flame arrester

Nitrogen
10 psig

split range
N₂ blanket
system set
at 500 Pa

PV valve
set at
−250 Pa
+750 Pa

Normal working:
ambient temp.
500 Pa

250 m³ intermediate
storage tank (IST)

C6 hydrocarbons
from reflux drum
on exit level
control (25 m³/hour)

seal pot

to petrol
blending system

J1 centrifugal transfer pump

*Figure A5.1 P&ID for the existing process.*

Figure A5.1 gives sufficient detail for the Procedural HAZOP and any deficiencies are outwith the scope of the study.

The modification is planned to remove the 100 m length of 100 mm diameter piping between the tank master isolation valve, V1, and the first pump isolation valve, V2, and to refit the pump closer to the tank, but first the flammable fluid (about 0.8 m$^3$) must be removed. Consideration was given to draining it into a drum but the risks were considered to be unacceptable. In line with the corporate management of change policy, a Hazards Study approach (see Chapter 2) was adopted.

1. The inherently safer option (HS 0) of displacing the fluids with nitrogen has been adopted. There is a nitrogen ring main on the site which can be connected below valve V4. This link is fitted with a non-return valve at the hose connection.
2. During the HS 2 analysis (FEED), it was recognized that this was a non-standard operation with potential *human factors* (see Section 10.2). It was recommended that all valves should be clearly labeled (V1, etc.) and that a dummy run practice should be carried out to debug the procedure and to familiarize the crew with the operation.
3. During the HS 3, it was decided that a HAZOP study should be carried out.

The final arrangement below (Figure A5.2) shows the 100 mm diameter suction line, the 25 mm diameter nitrogen header and a flexible hose, and 18 mm diameter depressuring line with an isolation valve V6. All piping other than the hose will be hard piped.

An operational procedure has been drawn up which, in accordance with the MOC policy, is to be the subject to a Procedural HAZOP.

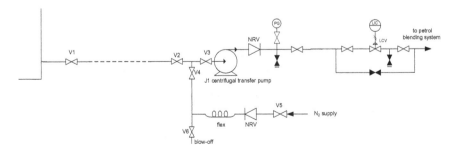

*Figure A5.2 Nitrogen supply connection for the line flushing procedure.*

## A5.2 DETAILED PROPOSED SEQUENCE

The operation will be carried out by an operator stationed near J1 who will be the lead operator and a second operator, in radio communication, at the tank to operate valve V1. The lead operator will control the procedure.

The initial set-up is for all valves V1−V6 closed and with the line between V1 and V2 containing $C_6$ liquid.

1. Open V6 then open V5 to prove line clear of debris and to displace any air in the hose.
2. Close V6 then open V4.
3. The operator at the J1 pump should open V2 slowly until fully open.
4. The operator at the tank is then instructed to open V1 slowly by one or two turns.
5. The operator at V1 should wait until nitrogen is heard passing through the valve into the IST then the tank operator will close V1.
6. The tank operator should then cautiously reopen V1 by one or two turns to ensure as much liquid as possible has been blown back to the IST.
7. Close V4.
8. Close V1 after allowing any residual $N_2$ in the line to depressure into IST.
9. Pump-based operator to close V2.
10. Close V5.
11. Verify V2, V4, and V5 are all closed.
12. Open V6 to depressurize the line.
13. Disconnect the hose at V4.

## A5.3 THE HAZOP STUDY

*HAZOP study team*

- Mike Manchester (MM) Facilitator and Scribe
- Brenda Bolton (BB) Production Manager
- Sandy Southport (SS) Senior Operator
- Wally Wigan (WW) Safety Officer

*Division into nodes*

Node 1

- Steps 1–2: Connect and prove the nitrogen supply. V5–V6.
- Design intention: To prove that the $N_2$ supply is fitted and to displace any air in the hose.

Node 2

- Steps 3–9: Clear the line by blowback to IST. V5–V1.
- Design intention: To completely clear petroleum feedstock from the 100 m line between the J1 pump and the IST by blowback to the IST using $N_2$ from the 1.3 barg nitrogen ring main via a temporary connection fitted to an existing drain by the J1 pump. Manual control by operators positioned at each end of the line. After the main clearance, a brief second flush will be applied.

Node 3

- Steps 10–13: Depressurize and disconnect. V1–V6.
- Design intention: Line previously containing $C_6$ but now containing $N_2$ to be depressured.

Node 4

- V3-J1-LCV: Line to petrol blending (not completed).

*Guide words*:

- Out of Sequence—too early, too late
- Rate—too fast, too slow
- Magnitude—more, less
- Pressure—more
- Communication
- Reverse
- Incomplete
- Other

# Table A5.1 HAZOP study report for node 1 (to be read in conjunction with figures A5.1 and A5.2)

Steps 1–2: Connect and prove the nitrogen supply. V5–V6.
Design Intention: To prove that the $N_2$ supply is fitted and to displace any air in the hose.
Initial status: All valves V1–V6 closed.
Attendees: MM, BB, SS, and WW (Note: BB* means BB is actioned to refer to a competent person).

Date: 2/1/15.

| Ref. No. | Guideword | Effect | Cause | Consequence | Safeguards | Actions | On |
|---|---|---|---|---|---|---|---|
| 1.1 | Out of sequence | No flow $N_2$ | V6 closed, V5 open | Some (small amount) of air left in the hose | Valve labeling and practice | 1.1.1 Reinforce practice<br>1.1.2 Confirm that traces of air in IST are not a safety issue | BB<br>WW |
| 1.2 | Too fast | High flow $N_2$ | V5 too far open | Waste of $N_2$ and local noise<br><br>Jet reaction on stones at vent, V6 | Valve labeling and practice | 1.2.1 Consider the jet reaction at the vent and secure<br>1.2.2 Can stones be sprayed about? Is there an "impact hazard" for humans?<br>1.2.3 Procedure to emphasize that the V5 should be opened slowly | BB<br>BB<br><br>WW |
| 1.3 | Out of sequence | Possible contamination of downstream process or rotation of the pump | V3 open by mistake, V6 closed, and V4 opened | Possible contamination of downstream process or rotation of the pump | Valve labeling and practice | 1.3 Consider locking V3 closed as part of the preparation process | BB |
| 1.4 | Out of sequence/ reverse | Reverse flow from (upstream) process to V6 | V3 left open, V4 left open, V6 open during the depressuring | Possible release of "materials" from downstream plant | Pump NRV<br>Valve labeling and practice<br>Many valves in route | Very low risk. See 1.3 above. Many valves have to be incorrectly set and NRV passing<br>Noted | No actions arising |
| 1.5 | Out of sequence/ reverse | Reverse flow from IST | V6 left open after blowdown and V1 and V2 opened ready for displacement of $C_6$. V5 still closed | IST drains through V6<br>Environmental impact<br>Potential fire | Valve labeling and practice | 1.5.1 Consider the need for NRV on the pump side of V2<br>1.5.2 This part of the procedure should have "one on one" supervision | BB*<br>BB |
| 1.6 | | Other guidewords | No effects identified | | | | |

## Table A5.2 HAZOP study report for node 2 (to be read in conjunction with figures A5.1 and A5.2)

Steps 3–9: Clear the line by blowback to IST. V5–V1.

Design intention: To completely clear petroleum feedstock from the 100 m line between the J1 pump and the IST by blowback to the IST using $N_2$ from the 1.3 barg nitrogen ring main via a temporary connection fitted to an existing drain by the J1 pump. Manual control by operators positioned at each end of the line. After the main clearance a brief second flush will be applied.

Status: As at end of Node 1.

Attendees: MM, BB, SS, and WW.

Date 2/1/15.

| Ref. No. | Guideword | Effect | Cause | Consequence | Safeguards | Actions | On |
|---|---|---|---|---|---|---|---|
| 2.1 | Out of sequence Too early (valve operation) | $N_2$ flow out of V6 | V6 left open | $N_2$ losses | Valve labeling and practice | 2.1.1. Obvious, take corrective action on V6 | BB* |
| | | Possible release of $C_6$ at V6 | Human factors. V6 opened and V1 and V2 opened for displacement. V5 not yet opened | Environmental impact and possible fire | Valve labeling and practice | 2.1.2. See 1.5.1 and 15.2 | BB |
| 2.2 | Too late (valve operation) | $N_2$ in next (downstream) operation | V2 not open, V3 left open | None | | 2.2 Action to be corrected See 1.3 | BB |
| 2.3 | Out of sequence | $C_6$ released from V6 | V1 and V2 open, V6 left open, and V5 closed | Environmental impact and possible fire | Valve labeling and practice | 2.3.1 Consider the need for an NRV at the $N_2$ side of V4 2.3.2 Review how this operation should be supervised. This part of the procedure should have "one on one" supervision. See 1.5.2 | BB WW |
| 2.4 | Too fast (valve opening) More flow pressure—more in IST Magnitude (more than two turns on V1) | V5 too far open More $N_2$ flow into IST | Human factors Poor understanding of operation Poor understanding of operation | Possible overpressure of IST due to high $N_2$ flow As above As Above | PRV on IST | 2.4.1 Assess the capacity of IST PRV against blow by 2.4.2 Consider the need for a flow restrictor in $N_2$ supply 2.4.3 If a flow restrictor is inserted how will it be controlled as it is now a "Safety Critical Item"? | BB* BB* BB* |

(Continued)

**Table A5.2 (Continued)**

| Ref. No. | Guideword | Effect | Cause | Consequence | Safeguards | Actions | On |
|---|---|---|---|---|---|---|---|
| 2.5 | Too slow (valve V5 opening) Low flow $N_2$ and $C_6$ | $C_6$ not displaced | No true indication of $N_2$ flow rate | Slower displacement of $C_6$ due to N2 "slippage"—wavy flow may result in limited $C_6$ removal | None | 2.5.1 The flow of $C_6$ will not necessarily be plug flow. In what two-phase flow regime is the displacement expected to operate? 2.5.2 How can the regime be controlled? | BB* <br><br><br><br> BB* |
| 2.6 | More flow $N_2$ High flow | See 2.4 and 2.5 | See 2.4 | See 2.4 | See 2.4 | 2.6.1 See 2.5.1 2.6.2 See 2.5.2 2.6.3 See 2.4.2/2.4.3 | BB* BB* BB* |
| 2.7 | Incomplete | $C_6$ left in line | Line not true, hogs and hollows plus elevation changes | Some $C_6$ trapped in the line at the end of the final blow through. Environmental impact and possible fire | None obvious | 2.7.1 Check the line slope and sags 2.7.2 Is there too much line distortion to make the blow out viable? A site visual check should be carried out | BB BB |
| 2.8 | Reverse flow of $C_6$ | $C_6$ released from V6 | V6 left open and V5 closed at the end of cycle. Some $C_6$ still in the line | Environmental impact and possible fire | Valve labeling and practice | As 2.3.1 | BB |
| 2.9 | Communication | As above <br><br> Misinterpretation | As above <br><br> Human factors What is the significance of a change in the noise? What will it sound like? | As above <br><br> Wavy flow may produce a sound like gas passing into the IST. | Valve labeling and practice | 2.9.1 Review how this operation should be supervised How long might it take? 2.9.2 Ensure the operators are trained in the use of radios 2.9.3 Review this parameter. Is it really safe for operation and a credible control parameter? | BB* <br><br> BB* <br><br> BB* |

(Continued)

# Table A5.2 (Continued)

| Ref. No. | Guideword | Effect | Cause | Consequence | Safeguards | Actions | On |
|---|---|---|---|---|---|---|---|
| 2.10 | Communication | As above | Misunderstanding of the point in the sequence without a clear lead operator | Possible upset not easy to define | Lead operator is specified in the procedure | 2.10.1 Ensure that one operator is clearly the lead operator controlling the actions and the other takes instructions from the leader | BB |
| | | | Misunderstanding of the point in the operation due to poor radio protocol | As above | | 2.10.2 Ensure that the operators are competent in the use of radios and language protocol | WW |
| | | Possible source of ignition | Radios not compatible with Hazardous Area Classification | Possible fire (remote possibility) | | 2.10.3 Verify that the radios are compatible with the area classification | WW |
| 2.11 | Incomplete | Possible reverse flow from IST during the step 8 depressuring cycle | Hydrostatic head in IST | Live V1–V2 is recontaminated with $C_6$ | None | 2.11.1 Consider closing V1 IMMEDIATELY the gas flow is detected and then depressure via V6 | BB |
| | | | | Possible environmental impact and fire during final blow down through V6 | | 2.11.2 Review the operation step 8 in the procedure. Is it viable? | BB |
| | Other guidewords | No effects identified | | | | | |

**Table A5.3 HAZOP study report for node 2, final blow through, steps 6–9 (to be read in conjunction with figures A5.1 and A5.2)**

(Immediate continuation of node 2 after first nitrogen flush of the line, i.e., completion of step 5.)
Status: As at the end of main blow through. V1, V3, and V6 closed; V2, V4, and V5 open.
Team/date as Table A5.2.

| Ref. No. | Guideword | Effect | Cause | Consequence | Safeguards | Actions | On |
|---|---|---|---|---|---|---|---|
| 2.12 | Less flow Incomplete | Line V1–V2 incompletely cleared. $C_6$ still in line | Flow regime uncertain and line slopes uncertain | Significant final $C_6$ left in line which has to be drained. Environmental impact and risk of fire | None | 2.7 (see 2.5.1 and 2.5.2; 2.6.1 and 2.6.2) | BB |
| 2.13 | More flow ($N_2$) (V5 too far open) (See 2.3) | V5 too far open | Human factors<br><br>Poor understanding of operation | Possible overpressure of IST | PRV on IST | 2.12.1 Assess the capacity of IST PRV against blow-by. See 2.4.1<br>2.12.2 Consider the need for a flow restrictor in $N_2$ line. See 2.4.2<br>2.12.3 If a flow restrictor is inserted how will it be controlled as it is now a "Safety Critical Item"? See 2.4.3 | BB*<br><br>BB*<br><br>BB* |
| 2.14 | Communication | $C_6$ still in line. Could result in a major spill later in the process | "Sound" is the only variable. What will it "sound like"? | Environmental impact and possible fire if incompletely drained | None | See 2.5 and 2.9.3<br>Review this parameter. Is it really safe for operation? | BB |
| | No other differences between first and final clearing | | | | | | |

## A5.4 FINAL HAZOP STUDY REPORT

Obviously, the final report cannot be written until the full HAZOP study has been completed. However, it is clear that there are a number of steps with potential for errors (human factors), it being a one-off operation and unfamiliar to the operations staff. There are also a number of "unknowns." Displacement (or blowing through) is a standard operation but it has significant implications when carried out within the constraints of this procedure. Is it a viable solution?

To date the key findings from nodes 1 and 2 are:

1. The procedure has missed the natural hogs and hollows in the line between V1 and V2 and any elevation changes between the nitrogen injection point and the IST which may make the procedure nonviable.
2. There is potential for release of $C_6$ at V6.
3. A possible overpressure of the IST following a $N_2$ blow-by.
4. Step 4 is vague—one or two turns is not a measurable parameter so there is potential for human error/factors.
5. The flow regime in the line from V2 to V1 is uncertain, and it is not clear that the contents can be displaced in a controlled manner. There is a potential conflict between transport of $C_6$ and IST integrity. Is there a better alternative?
6. The compatibility of the radios with the Hazardous Area Classification.
7. It is a new one-off operation which needs some training (human factors).
8. There is a need for labeling of all the valves (human factors).
9. The interpretation of the "end of clearing" using a subjective "noise" (human factors) is not inherently safe.
10. There is a possibility of recontamination of the line following the final blow out due to hydrostatic head in IST if the closure of V1 is delayed. The line V1−V2 must not be blown down into IST due to the risk of recontamination. V1 must be closed first.

## A5.5 AUTHORS NOTES ON THIS PROCEDURE

Valve V1 is a gate valve as the procedure says "open two turns." Assuming that it is mounted vertically, the flow regime in the line from V1 to V2 will be very uncertain as the gap for liquid flow will be at the bottom of the line, and this will not necessarily be the low point—there could be "hogs and hollows" in the line especially if it slopes to the pump J1. A foam pig would pass the fully open gate valve V2 and be stopped by the partially open gate valve V1 (open one or two turns). This method would give a more complete line clearance and so a pig run may be preferable.

The objective of this exercise was to demonstrate the use of HAZOP in a procedure, but it has produced more issues than expected! This shows the strength of HAZOP.

| | |
|---|---|
| AFD | approved for design |
| AIChE | American Institution of Chemical Engineers |
| BS | British Standard |
| CFR | Code of Federal Regulations |
| CHAZOP | computer HAZOP (study) |
| CIA | Chemical Industries Association |
| CPF | central process facility |
| DHSV | down hole safety valve |
| EC | European Community |
| EPSC | European Process Safety Centre |
| ESDV | emergency shutdown valve |
| EU | European Union |
| FEED | front-end engineering design |
| FMEA | failure modes and effects analysis |
| FMECA | failure modes and effects criticality analysis |
| HAZID | hazard identification (method) |
| HAZOP | hazard and operability (study) |
| HS | hazard study |
| HSE | Health and Safety Executive (UK) |
| IChemE | Institution of Chemical Engineers |
| IEC | International Electrotechnical Commission |
| I/O | input/output |
| LOPA | layer of protection analysis |
| LTEL | long-term exposure limit |
| MOC | management of change (system) |
| MSDS | material safety data sheet |
| MV | master valve |
| OP | orifice plate |
| OSHA | Office of Safety and Health Administration (USA) |
| P&ID | piping and instrumentation diagram |
| PES | programmable electronic system |
| PFD | probability of failure on demand |
| PI | productivity index |
| PIF | performance-influencing factor |

| PSM | process safety management |
| PSSR | pre start-up safety review |
| QA | quality assurance |
| QRA | quantitative risk assessment/analysis |
| SHE | safety, health, and environmental |
| SIL | safety integrity level |
| SIS | safety instrumented system |
| SMS | safety management system |
| SOP | standard operating procedure |
| SOR | safety and operability review |
| SSSV | sub-surface safety valve |
| WV | wing valve |

# REFERENCES AND BIBLIOGRAPHY

1. Kletz T. *HAZOP and HAZAN*. 4th ed. Rugby, UK: IChemE; 2006.

2. Crawley FK, Tyler BJ. *Hazard identification methods*. Rugby, UK: EPSC/IChemE; 2003.

3. AIChE. *Guidelines for hazard evaluation procedures*. 3rd ed. USA: AIChE Center for Chemical Process Safety; 2008.

4. Gillett JE. *Hazard study and risk assessment in the pharmaceutical industry*. IL: Interpharm Press; 1997.

5. Wells G. *Hazard identification and risk assessment*. Rugby, UK: IChemE; 2005.

6. ISSA Prevention Series No 2002, IVSS-Sektion Chemie, Heidelberg, Germany. Revised edition of *Risikobegrenzung in der Chemie*. PAAG-Verfahren (HAZOP).

7. BS EN 61882:2001 Hazard and operability studies (HAZOP studies). Application guide (*IEC 61882, Guide for Hazard and Operability (HAZOP) Studies*).

8. Jones D. *Nomenclature for hazard and risk assessment in the process industries*. 2nd ed. Rugby, UK: IChemE; 1992.

9. See the HSE website <www.hse.gov.uk>. (Note that Seveso III is due in 2015.)

10. See the OSHA website <www.osha.gov/law-regs.html>.

11. Swann CD, Preston ML. Twenty-five years of HAZOPs. *J Loss Prev Process Ind* 1995; **8**(6):349–53.

12. Kletz T, Amyotte P. *Process plants—a handbook for inherently safer design*. 2nd ed. USA: CRC Press; 2010.

13. AIChE. *Inherently safer chemical processes—a lifecycle approach*. 2nd ed. USA: AIChE Center for Chemical Process Safety; 2006.

14. BS EN 61511 Functional safety. Safety instrumented systems for the process industry sector.

15. AIChE. *Layer of protection analysis: simplified process risk assessment*. USA: AIChE Center for Chemical Process Safety; 2001.

16. Pitblado R, Turney R. *Risk assessment in the process industries*. 2nd ed. Rugby, UK: IChemE; 1996.

17. HSE. *Quality assurance of HAZOP*. Sheffield: HSE Information Services; 1996 *HSE offshore technology report, OTO 96 002*.

18. AIChE. *Guidelines for auditing process safety management systems*. 2nd ed. USA: AIChE Center for Chemical Process Safety; 2011.

19. Health and Safety Executive. The explosion and fires at the Texaco Refinery, Milford Haven, 24 July 1994.

20. Tyler BJ. HAZOP study training from the 1970s to today. *Process Saf Environ Protect* 2012;**90**:419.

21. *Out of control: Why control systems go wrong and how to prevent failure*. HSE Books; 2003.

22. Kletz T. *Computer control and human error*. Rugby, UK: IChemE; 1995.

23. BS IEC 61508 Functional safety of electrical/electronic/programmable electronic safety-related systems, 1999.

24. *Programmable electronic systems in safety related applications. Part 1 An introductory guide, Part 2 General Technical Guidelines.* HSE Books; 1987.

25. Andow P. *Guidance on HAZOP procedures for computer controlled plant.* UK: HSE; 1991 *HSE research report no 26/1991.*

26. *HSG48 Reducing error and influencing behavior.* HSE; 1999.

27. See the HSE website <http://www.hse.gov.uk/humanfactors>.

28. AIChE. *Guidelines for preventing human error in process safety.* New York, NY: AIChE Center for Chemical Process Safety; 2004.

29. *Second report of the study group on human factors.* London: HSC; 1991. ISBN 0 11 885695 2.

30. Embrey DE. Quantitative and qualitative prediction of human error in safety assessments. *Major hazards onshore and offshore.* IChemE Symp. Ser. No. 103; 1992.

31. Identifying Human Failure, Core topic 3 in human factors: inspectors human factors toolkit. At <http://www.hse.gov.uk/humanfactors/toolkit.htm>.

32. *Layer of protection analysis: simplified risk assessment.* AIChE Center for Chemical Process Safety; 2001. ISBN 978 0 8169 0811 0.

33. *Buncefield, Safety and environmental standards for fuel storage sites.* Process Safety Leadership Group, Final report 2009, HSE, ISBN 978-0-7176-6386-6. See Appendix 2: Guidance on the application of LOPA to the overflow of a gasoline storage tank operated at atmospheric pressure.

34. Barton J, Rogers R. *Chemical reaction hazards.* 2nd ed. Rugby, UK: IChemE; 1997.

# BIBLIOGRAPHY

EPSC. *Safety management systems.* Rugby, UK: IChemE; 1994.

IVSS. 1999, Das PAAG-Verfahren. Methodik, Anwendung, Beispiele.

Knowlton RE. *A manual of hazard and operability studies.* Chemetics International Co Ltd; 1992.

Lees FP. 4th ed. Mannan S, editor. *Loss prevention in the process industries,* vols. 1–3. UK: Butterworth–Heinemann; 2012.

Skelton B. *Process safety analysis: an introduction.* Rugby, UK: IChemE; 1997.

*Note*: Page numbers followed by "*f*" and "*t*" refers to figures and tables respectively.

Printed in the United States
By Bookmasters